INVENTING THE
FUTURE

Also by F. Clifton Berry, Jr.:
Sky Soldiers (1987)
Strike Aircraft (1988)
Chargers (1988)
Gadget Warfare (1988)
Air Cav (1988)

With Tom Allen and Norman Polmar:
CNN: War in the Gulf (1991)

INVENTING THE FUTURE

How Science and Technology Transform Our World

F. Clifton Berry, Jr.

BRASSEY'S (US), Inc.
A Division of Maxwell Macmillan, Inc.

Washington • New York • London

Cover design and graphics: Clint Chadbourne
Book design: Mary W. Matthews

Brassey's (US), Inc.

Editorial Offices
Brassey's (US), Inc.
8000 Westpark Drive
First Floor
McLean, Virginia 22102

Order Department
Brassey's Book Orders
c/o Macmillan Publishing Co.
100 Front Street, Box 500
Riverside, New Jersey 08075

Brassey's (US), Inc., books are available at special discounts for bulk
purchases for sales promotions, premium, fund-raising, or educational
use through the Special Sales Director, Macmillan Publishing Company,
866 Third Avenue, New York, New York 10022.

Library of Congress Cataloging-in-Publication Data:
Berry, F. Clifton, Jr.
        Inventing the future : how science and technology transform our
world / F. Clifton Berry, Jr.
            p. cm.
        Includes bibliographical references and index.
        ISBN 0-02-881029-5
        1. Electronics—United States—Technological innovations.
        2. Telecommunication—United States—Technological innovations.
        3. Inventors—United States.   I. Title.
        TK7811.B47   1992   92-37423
        621.381'0973—dc20   CIP

10 9 8 7 6 5 4 3 2 1

Printed in the United States of America

To the men and women scientists and engineers whose work has transformed our world since 1950, and to the young men and women who will invent the future of the twenty-first century.

Trademarks

Every effort has been made to acknowledge the trademark status of companies and products mentioned in this book. Company and product names not included in the following list are considered to be trademarks or registered trademarks of their respective companies.

Allied Chemical is a registered trademark of Allied Chemical Corp. Apple and Apple II are registered trademarks of Apple Computer, Inc. AT&T is a registered trademark of American Telephone and Telegraph Co. Atari is a registered trademark of Atari, Inc. Battelle is a registered trademark of Battelle Memorial Institute. BBN is a registered trademark of Bolt, Beranek, and Newman, Inc. Bellcore is a registered trademark of Bellcore Corp. BellSouth is a registered trademark of BellSouth Corp. Bendix is a registered trademark of The Bendix Corp. Boeing is a registered trademark of The Boeing Co. Byte Shop is a registered trademark of Byte, Inc. Centralab is a registered trademark of Centralab, Inc. CNN is a registered trademark of Cable News Network. COMSAT is a registered trademark of Communications Satellite Corporation. Corning is a registered trademark of Corning, Inc. Encyclopedia Britannica is a registered trademark of Encyclopedia Britannica, Inc. Federal Express is a registered trademark of Federal Express Corp. Ferranti is a registered trademark of Ferranti Electric, Inc. Fuzzbuster is a registered trademark of Dale T. Smith. General Dynamics is a registered trademark of General Dynamics Corp. General Electric is a registered trademark of General Electric Co. General Motors is a registered trademark of General Motors Corp. Gentex is a registered trademark of Gentex Corp. GRiD and PalmPad are trademarks of GRiD Systems Corp. Grumman is a registered trademark of Grumman Corp. Hazeltine is a registered trademark of Hazeltine Corp. Heinz is a registered trademark of H.J. Heinz Co. Hewlett-Packard is a registered trademark of Hewlett-Packard Co. Hughes is a registered trademark of Hughes Aircraft Co. IBM and IBM PC are registered trademarks of International Business Machines Co. IMM is a registered trademark of International Mobile Machines Corp. Intel is a registered trademark and 4004, 8008, and 8088 are trademarks of Intel Corp. INTELSAT is a registered trademark of International Telecommunications Satellite Organization. ITT and GILFILLAN are registered trademarks of ITT, Inc. The JASON Project is a registered trademark of The JASON Foundation for Education. L.L. Bean is a registered trademark of L.L. Bean, Inc. Lands' End is a registered trademark of Lands' End, Inc. Lockheed is a registered trademark of Lockheed Aircraft Corp. Marconi is a registered trademark of Marconi's Wireless Telegraph Company Ltd. Martin Marietta is a registered trademark of Martin Marietta Corp. Microsoft and MS-DOS are registered trademarks of Microsoft Corp. Motorola is a registered trademark of Motorola, Inc. National Inventors Hall of Fame is a registered trademark of The National Council of Patent Law Associations. *Nature* is a registered trademark of Macmillan Journals Ltd. NBC is a registered trademark of the National Broadcasting Company, Inc. Nobel Prize is a registered trademark of The Nobel Foundation. Northrop is a registered trademark of Northrop Corp. Perkin-Elmer is a registered trademark of Perkin-Elmer Corp. Philco is a registered trademark of Philco International Corp. Photophone is a registered trademark of RCA Incorporated and Image Data Corp. PictureTel is a trademark of PictureTel, Inc. Raytheon is a registered trademark of Raytheon Co. Rolls-Royce is a registered trademark of Rolls-Royce, Ltd. Spectra-Physics is a registered trademark of Spectra-Physics Scanning Systems, Inc. Sperry is a registered trademark of Sperry Rand Corp. Texas Instruments is a registered trademark of Texas Instruments, Inc. Thomson-CSF is a registered trademark of Thomson-CSF, Inc. TracView is a registered trademark of Hughes Aircraft Co. TRW is a registered trademark of Thompson, Ramo, Wooldridge, Inc. TV Answer is a registered trademark of TV Answer Inc. Union Carbide is a registered trademark of Union Carbide Corp. Volkswagen is a registered trademark of Volkswagenwerk Aktiengesellschaft. Wal-Mart is a registered trademark of Wal-Mart Stores Inc. Westinghouse is a registered trademark of Westinghouse Electric Corp.

Credits

All photos are from Hughes Aircraft, except as noted below.

All photos are used with permission, and the author and publisher acknowledge with thanks the granting of permission for their use. Names of sources, listed alphabetically, are followed by the page number(s) of their photo(s).

A1C Andy Rice, Combat Photographer: 31; Air Force Association: 48, 75, 118, 144; Apple Computer: 28; AT&T Information Services Network: 60; Battelle Today: 57; BBN Systems and Technologies: 145; Bellcore Labs: 67; Corning, Inc.: 64, 65; Dept. of Defense (DoD): 86, 108, 112, 113, 117, 126, 148, 149; FAA: 76; GRiD Systems: 27; Grumman: 120; IBM: 19, 94; IDEO/David Kelley Design: 29; Intel Corp.: 13, 24; INTELSAT: 37, 40; International Mobile Machines: 134; JASON Foundation: 153; Microsoft: 30; Motorola: 26; NASA: 96, 98, 101, 102, 105; Oaknoll Medical Group: 96; PC Glossary, Disston Ridge: 25; PictureTel: 140; Smithsonian Institution: 20, 21, 22, 23; Spectra-Physics: 56; T.V. Answer: 138; Texas Instruments: 2, 7, 9, 10, 12, 14, 16; Tracey Attlee: 18; U.S. Air Force: 123; U.S. Dept. of the Interior, Bureau of Mines: 5, 6; United States Surgical Corporation: 159; Westinghouse: 70.

Contents

Foreword

Since 1950, the rapid advances of technologies have invented the future even as we lived the present. The new technologies have profoundly transformed society and have created "one world."

Nowadays, we take the swift march of technology for granted. We have become accustomed to innovation, and expect it as a matter of course. Tomorrow, innovation will continue to be a way of life, with technology advances cascading into the marketplace at ever-increasing speed.

Developments in the four decades since 1950 surpass the "golden ages" of the past. The technological renaissance that gathered momentum in the 1950s and continues in the 1990s is exemplified by the field of electronics. Hughes Aircraft Company was at its heart, and it continues to be a world leader in scientific and technological innovation.

Over a span of more than four decades, the period covered in *Inventing the Future*, the men and women of our industry maintained an unrelenting quest for new concepts and advances that can "make a difference." Their continuous advancement of the frontiers of science and technology brought great personal and professional satisfactions. Both of us, who have been privileged to lead within this industry, have seen in our own careers the beginnings of great technological breakthroughs and have known the men and women who advanced the frontiers. We know the excitement that crackles through a company and its people when new things are created, from communications satellites and lasers to personal computers.

That excitement continues in the mid-1990s and will endure. Because innovations beget more innovations, the opportunities for young men and women will be multiplied beyond imagining.

Reflecting upon the rapid march of technology in our own careers, we believe that society must come to realize that technology enables humankind to shape the future for good or ill. By proper use of technology, we can spur economic growth, raise standards of living, and improve the quality of life in our cities and on this planet. Or, through improper use of the power of technology, we can turn in the wrong direction.

Thus the important question: Can we work together — government, industry, and academia — in new partnerships that will extend the benefits of technology to all?

That is an open question. Searching for its answers will be as exciting and rewarding for the world as are the technologies chronicled here.

C. Michael Armstrong
Chairman and CEO
Hughes Aircraft Company

Dr. Malcolm R. Currie
Chairman Emeritus
Hughes Aircraft Company

Los Angeles, California

Preface

This book is intended as a contemporary reference on key electronics technologies and their impact on our world. It is not a textbook that reveals the equations underlying the technologies. Nor is it a set of biographies of the key inventors of the enabling technologies chronicled here.

Rather, *Inventing the Future* highlights several of the crucial technologies invented in the United States in the past four decades, and tells how they were integrated into applications that wrought unprecedented transformations of our world. Perhaps this telling will stimulate men and women of all ages who are seeking challenge and satisfaction in their lives to contemplate careers in science and technology. For what is more challenging and satisfying than advancing knowledge in ways that fellow humans can use, and thereby improve their lives?

Finally, this book is aimed at satisfying the curiosity of a wider audience, people who are intrigued by technology and want to know how it came to be and how it affects their lives.

Over the past four decades, science and technology have transformed our world in ways that surpass the fantasies of the wildest science fiction.

In the brief span since 1950, especially in the United States, we have witnessed more advances in technology and greater change in the lives of people than in all of previous recorded history.

Thanks to the new technologies, we live in a world without frontiers. It is a world where ideas and information and data and pictures are spread everywhere instantaneously. Electronics in the broadest sense has been a major enabling factor in this global transformation. Electronics and associated technologies have changed humanity's way of life so profoundly that its marvels are taken for granted in advanced societies.

The changes of the past four decades are extraordinary enough to cause wonderment. Yet, as humans stumble toward the twenty-first century, the wonders of technology have only begun. The future is being invented every day.

Many experts on the technologies covered here have seen parts of the manuscript, and I am grateful for their careful review. Any errors that may be in the text are not their doing, and are solely my responsibility.

F. Clifton Berry, Jr.

Part I
Enabling Technologies

They are the foundations for
transforming our world

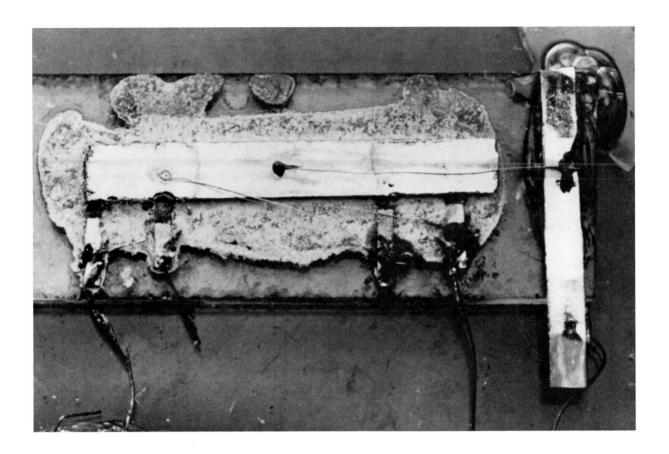

Jack S. Kilby's first integrated circuit was a little more than 11 mm (7/16 inch) long and contained the equivalent of five electronic devices — one transistor, three resistors, and one capacitor.

Chapter 1
Enabling the Information Age

The integrated circuit was the key to the electronic revolution of the second half of the twentieth century, and to the dawning of the Information Age.

Humans crave information, wanting to know more about themselves and their world. Over the past four decades, breakthroughs in electronic technology have made possible new and unexpected ways of gathering, processing, and distributing information to satisfy that human craving. Society, in turn, has found ways to use new information faster to cope more effectively.

Alvin Toffler, in his 1970 book *Future Shock,* wrote about how change would accelerate in the latter third of the twentieth century. Focusing on the advanced nations of the world, he wrote:

"... the child reaching teen age in any of these societies is literally surrounded by twice as much of everything newly man-made as his parents were at the time he was an infant."

By the time today's teenager reaches 30, a second doubling will have occurred. Within a 70-year lifetime, perhaps five such doublings will take place.

Toffler saw technology as a major force behind the acceleration of change. He identified three stages in technological innovation: the creative, feasible idea; its practical application; and its diffusion through society. The acceleration of change occurs because technology feeds on itself and makes more technology possible.

Electronics has spawned many of the technologies that transformed the world in the past four decades. In half a lifetime, electronics has opened doors to rich and limitless sources of information.

This book concentrates on inventions and applications in electronics since about 1950. "Electronics" is interpreted broadly as ranging across the entire electromagnetic spectrum.

The last 40 years have been a period of innovation unprecedented in history, a brief era that has seen an outpouring of inventions and innovations that have laid foundations for even more extraordinary developments. The developments in electronics have expedited discoveries and innovations in fields as diverse as medicine, biology, geology, and archeology. Thousands of men and women were responsible for the rapid advances. They worked in hundreds of companies, large and small, in academic laboratories and in their basement or garage workshops. They came from all parts of the country and from all backgrounds. Their brain power produced the integrated circuit, the microprocessor, the geosynchronous communications satellite, lasers, fiber optics, radar, and thermal imaging.

The people who are a central focus of the book share certain significant characteristics, including persistence, an itch to explore the unknown, resistance to conventional wisdom, particularly when it says something is impossible, and a willingness to try multiple approaches to a challenge. When a direct route to a solution seems impossible, they find a way to go around the obstacles.

Looking back in the history of technology, it is apparent that these attributes are shared by innovators of the past, from Leonardo da Vinci and Galileo to Thomas Edison and Alexander Graham Bell. What is now different from the past, however, is that people in developed societies expect a continuous rush of inventions and new developments because of the technology explosion.

The wondrous discoveries and developments since the 1950s have been made possible through transforming of commonplace elements such as silicon and germanium into powerful levers that expand humanity's opportunities for discovery. The alchemists of the Middle Ages tried to transform lead, a base metal, into gold, the most precious of metals. They wanted to convert something cheap into something very valuable. In modern times, the achievements of science and technology surpass the most fanciful legends and become the new alchemy. For example, silicon, one of the commonest elements on Earth, has been transformed into a commodity that in many respects is far more valuable than gold or diamonds. Silicon is the basis for creating the integrated circuit, which is the enabling technology that makes other wonders of the modern age possible.

Silicon makes up more than one-quarter of the Earth's crust. After oxygen, it is the second most abundant element. In the form of silica and silicates it constitutes the bulk of most common rocks, sands, soils, and clays. It is virtually everywhere on Earth. Silicon's widest use is in the production of iron and steel. It is also used in quantity for aluminum alloys and is employed by the chemical industry as a raw material in numerous processes.

Silicon production in the United States in 1990 was valued at $390 million. That is less than half the value of dog food produced in the United States. A small fraction of that silicon production, perhaps $20 million worth, qualifies as semiconductor grade.

That tiny amount of silicon is the source of the New Alchemy. When purified, it becomes the fundamental building block for the semiconductor industry. After it is processed and transformed into semiconductors and microprocessors, the $20 million worth of raw material becomes the foundation for a $500 billion industry. That's transmutation on a grand scale.

Silicon, in fact, has been valuable to us since antiquity. The first tools and weapons were made from flint, a variety of quartz or silicon dioxide. According to experts at the Bureau of Mines, the use of

A silica mining operation near Ottawa, Illinois. Silica, after it is processed, is the basic material for integrated circuits.

silicates for pottery dates back to our earliest history, and its use for glass can be traced back to about 12,000 B.C. In the late nineteenth century, silicon's use in steel-making expanded, and silicon carbide (carborundum) was discovered, increasing the demand for silicon.

In modern times, researchers discovered that silicon's crystal structure makes it a very good semiconductor. In 1949, the E.I. duPont de Nemours company produced the first silicon pure enough for use in transistors and other semiconductor devices. A semiconductor can either conduct a flow of current, or it can resist the flow, depending on how the material is treated. One area of a slice of semiconductor material (such as silicon) can be chemically treated to be a good conductor, while another area can be treated to resist the flow of current. This enables current to flow through a circuit ("on") or conversely, to not pass ("off"). The result is the on-and-off switching that is essential for the functioning of solid-state devices such as transistors, and for the fundamental language of computers.

The crystal structure of silicon is diamond cubic, where each atom in the lattice shares its four outermost electrons (i.e., valence electrons) with the four nearest neighboring atoms through covalent bonding.

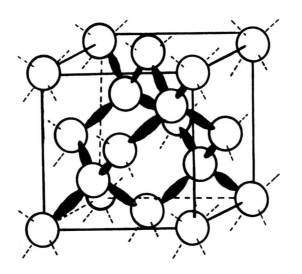

Silicon enjoys an edge over other semiconductor materials such as germanium because it is cheaper and its crystal structure is "diamond cubic." That is, each atom in its lattice structure shares its four outermost electrons with the four nearest neighboring atoms through a phenomenon called covalent bonding. This diamond cubic structure is a key feature of silicon's success as a semiconductor.

Role of Semiconductors

Semiconductors literally freed humanity from the shackles of mechanical linkages and the limitations of early electronics. From its earliest harnessing of current, electricity created labor-saving potential, to power machinery or produce light with the flick of a switch. But before the electronic age arrived, society had to depend on mechanically operated tools and devices.

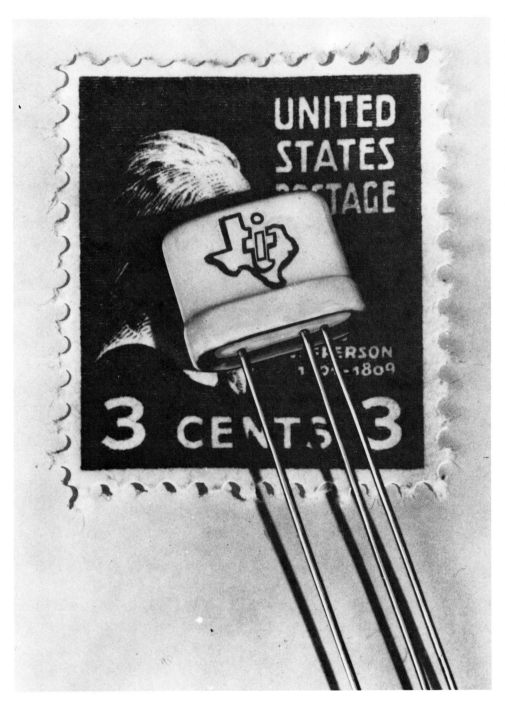

Texas Instruments began the first commercial production of silicon transistors in 1954. Many of these transistors were no larger than the average postage stamp.

The abacus is a primitive but still serviceable example of a mechanical calculator. The beads on the wires are manipulated mechanically to devise solutions. Other, more complicated, calculators consisted of collections of levers, cogs, and rods. Herman Hollerith's tabulating machine, used first in the 1890 Census, harnessed electricity to do some of the work, but at heart it was still mechanical. Its operations were lengthy and labor intensive. A new method was needed to calculate larger amounts of numbers in less time.

The invention of the vacuum tube in 1904 and its development into rectifiers, amplifiers, and other tools of electronics made faster calculators possible. But the vacuum tubes had to be connected by wires and were fragile, bulky, and relatively unreliable. They were essentially glass cylinders three or four inches high and an inch or more in diameter, with circuit elements visible through the clear glass and connector prongs protruding from their bases.

Vacuum tubes required a lot of power. They overheated and burned out, and did not last very long. But they were better than what had gone before. So vacuum tubes were exploited to the maximum, especially to meet the demands of World War II. For example, the Boeing B-29 Superfortress bomber, the most advanced U.S. strategic aircraft at the end of the war, used almost 1,000 vacuum tubes and tens of thousands of other electronic components in its radio, radar, and other electronic equipment.

In the early postwar years, vacuum tubes were the working heart of civilian radio and television receivers. Owners of radio and television sets became accustomed to buying vacuum tube replacements at drug and hardware stores to keep their equipment working.

The transistor overcame the limitations of vacuum tubes. When William Shockley and his colleagues John Bardeen and Walter Brattain invented the transistor in 1948, they created something entirely new. The transistor was a device that worked like a vacuum tube, but at the same time was more reliable, consumed less power, and was many multiples smaller. Large commercial transistors of the early 1950s were smaller than a three-cent postage stamp. Transistors soon found applications in everything from portable radios to hearing aids, and, as their uses expanded, the days of the vacuum tube became numbered.

Like vacuum tubes, however, transistors had to be connected in circuits in order to function. As transistors grew smaller and more applications for them were devised, the circuits became more complex and the transistors were connected in larger numbers. The connections involved thousands of soldering operations, performed by patient workers peering through huge magnifying glasses to manipulate hot soldering irons.

The very act of making the connections of greater numbers of transistors led to enormous problems. The more numerous the connections, the more labor was required and the more likely at least one or more connections would be defective or would fail in use. In the 1950s, engineers were thwarted by these problems.

The challenge lay in making the components and connections of electronic circuits in ways that were not only cost-effective but also reliable, and at the same time, making them in large numbers. The interrelationships of reliability, cost, and complexity created the Tyranny of Numbers, a tyranny that threatened to limit the applications of transistors.

Enter Jack Kilby

Jack St. Clair Kilby found the solution to the Tyranny of Numbers: the integrated circuit. Kilby was born in Jefferson City, Missouri in 1923, and grew up in Kansas with an interest in radio. Like so many young men of the era, he built his own radio equipment and experienced the satisfaction of capturing distant voices from the air and hearing them on his own receiver. He served in the Army Signal Corps during World War II and after the war earned bachelors and masters degrees in electrical engineering at the University of

A page from Jack S. Kilby's laboratory notebook, documenting the first working integrated circuit.

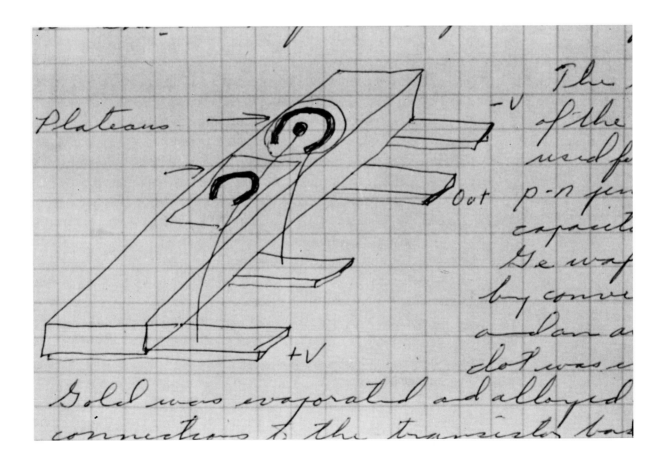

Jack S. Kilby, co-inventor of the integrated circuit, displays his invention.

Illinois in Champaign. His first job after the university was with a company called Centralab at Milwaukee, where he went to work in 1947.

Bell Labs held the transistor patent, but licensed it to others such as Texas Instruments and Centralab. By 1951, Centralab was exploiting the new technology to create ever-smaller hearing aids. In fact, the first commercial product using transistors was a hearing aid. Kilby designed and developed silk-screen circuits for the devices, and during his 11 years at Centralab he earned a reputation as an innovative electrical engineer.

By 1958, Kilby felt he had reached a career ceiling. Centralab was too small to underwrite the work in applied technology work he wanted to do. So he sent out résumés to larger companies in the electronics industry. He did not have to wait long for a response. Texas Instruments in Dallas wanted him. He joined a TI laboratory in the summer of 1958, just before the entire crew took their summer vacations. Since he had just arrived and had built up no vacation time, Kilby had the laboratory to himself. He knew the desirability of integrating all the elements of a circuit on a single base and was well aware of the problems of interconnecting the components, the implacable Tyranny of Numbers.

Now, in the hot summer days and nights in the nearly-empty laboratory in Texas, Kilby was about to overthrow that tyranny.

The Monolithic Idea

The idea for one monolithic circuit with all components on a single surface came to Jack Kilby during his solitary summer. He recorded the ideas in his engineering notebook, along with sketches of how it might be done:

> . . . I began to feel that the only thing a semiconductor house could make in a cost-effective way was a semiconductor.

Further thought led me to the conclusion that semiconductors were all that were really required; that resistors and capacitors (passive devices), in particular, could be made from the same material as the active devices (transistors). I also realized that, since all of the components could be made of a single material, they could also be made *in situ,* interconnected to form a complete circuit. I then quickly sketched a proposed design for a flip-flop using these components. Resistors were provided by bulk effect in the silicon, and capacitors by p-n junctions.

When Willis Adcock, Kilby's supervisor, returned from vacation, Kilby outlined his idea. Adcock wanted proof that it would work, so Kilby first built a simple multivibrator circuit using discrete silicon elements. He demonstrated it to Adcock on August 28, 1958. That was the proof that Adcock wanted, but it still was not an integrated circuit, so Kilby proceeded to build the first one. Because germanium was at hand, he used a wafer of it to build a phase-shift oscillator that contained a resistor, capacitor, and transistor, all created and interconnected from the same chunk of germanium.

On September 12, 1958, Kilby faced a crowd of executives and fellow engineers gathered around his setup in the laboratory. They wondered if the device would work as advertised. Kilby already knew it would. Leaving nothing to chance, he had tested the circuit on the previous day. As the others watched and hoped, Kilby applied a 10-volt current to the model integrated circuit hooked up to a cathode ray tube for display. The onlookers were treated to a perfect sine wave oscillating across the tube.

The integrated circuit was born.

Bob Noyce's Contribution

At roughly the same time, in the San Francisco Bay area, the idea of multiple electronic components on a chip of a single substance very much occupied the mind of Robert N. Noyce, one of the founders of Fairchild Semiconductor Co. His ideas were about to create another breakthrough in the quest for an integrated circuit.

When he was young, Noyce, a native of Burlington, Iowa, built model airplanes and a glider and, like Jack Kilby, ham radio sets. He earned his Ph.D. in Physical Electronics at the Massachusetts Institute of Technology in 1953. After three years with Philco, Noyce moved west to join William Shockley in Palo Alto, California. Shockley,

N°   3,138,743

THE UNITED STATES OF AMERICA

TO ALL TO WHOM THESE PRESENTS SHALL COME:

Whereas Jack S. Kilby, of Dallas, Texas, assignor to Texas Instruments Incorporated, of Dallas, Texas, a corporation of Delaware,

PRESENTED TO THE **Commissioner of Patents** A PETITION PRAYING FOR THE GRANT OF LETTERS PATENT FOR AN ALLEGED NEW AND USEFUL INVENTION THE TITLE AND A DESCRIPTION OF WHICH ARE CONTAINED IN THE SPECIFICATION OF WHICH A COPY IS HEREUNTO ANNEXED AND MADE A PART HEREOF, AND COMPLIED WITH THE VARIOUS REQUIREMENTS OF LAW IN SUCH CASES MADE AND PROVIDED, AND

Whereas UPON DUE EXAMINATION MADE THE SAID CLAIMANT is ADJUDGED TO BE JUSTLY ENTITLED TO A PATENT UNDER THE LAW.

NOW THEREFORE THESE **Letters Patent** ARE TO GRANT UNTO THE SAID

Texas Instruments Incorporated, its successors
OR ASSIGNS

THE TERM OF SEVENTEEN YEARS FROM THE DATE OF THIS GRANT

RIGHT TO EXCLUDE OTHERS FROM MAKING, USING OR SELLING THE SAID INVEN- THROUGHOUT THE UNITED STATES.

In testimony whereof, I have hereunto set my hand and caused the seal of the Patent Office to be affixed at the City of Washington this twenty-third *day of* June, *in the year of our Lord one thousand nine hundred and* sixty-four, *and of the Independence of the United States of America the one hundred and* eighty-eighth.

Attest:

Attesting Officer.

Commissioner of Patents.

one of the inventors of the transistor and a Nobel Laureate, had left Bell Labs and founded Shockley Semiconductor Laboratory in 1955. The opportunity to work with Shockley on interesting semiconductor projects attracted Noyce. But in 1957 Noyce and seven other scientists left Shockley to found Fairchild Semiconductor, where Noyce was first among equals.

By 1958 Fairchild Semiconductor was producing advanced transistors and doing very well, but Noyce was drawn to the challenge of over-

Robert N. Noyce, founder of Intel and co-inventor of the integrated circuit. After years of discussions, his company and Texas Instruments agreed that he and Jack Kilby were co-inventors.

coming the Tyranny of Numbers. He approached the solution from a different direction than Jack Kilby.

Noyce's breakthrough came in inventing a method of interconnecting the circuit elements in a semiconductor structure by superimposing a thin film of metal atop a wafer of silicon. Noyce and his colleagues at Fairchild Semiconductor had already improved the reliability of transistors by coating the transistor wafer surface with a layer of silicon dioxide. This prevented the phenomenon called surface inversion, which allowed current to leak across the transistor junctions. Silicon was chosen because its properties were well known and the processes of working with it were well established. Indeed, silicon was to supplant germanium in the manufacture of semiconductor devices and integrated circuits.

Fairchild used the process to make the world's most reliable transistors. That led Noyce to conceive the idea for the integrated circuit. He later recalled:

> The integrated circuit came out of my own laziness. We took those transistors that were nicely arranged on a piece of silicon, cut them in tiny little pieces . . . shipped them to the customers, and they put them all back together again. Why not cut out all the middle ground and just put them together while still on the silicon? That's what we did.

Opposite: Kilby's actual patent #3,138,743 for the integrated circuit, which he assigned to Texas Instruments.

Only one centimeter square, the Texas Instruments 32-bit LISP processor of 1988 crammed ever more power into an ever smaller package.

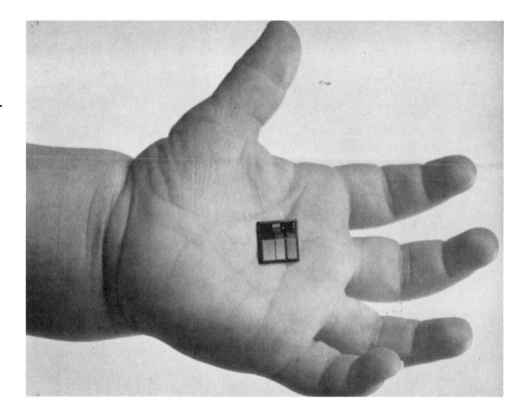

Noyce made mass production of integrated circuits possible.

Both Kilby and Noyce applied for and received patents on their creations. Texas Instruments filed an interference lawsuit against Fairchild Semiconductor over Noyce's patent. A long period of litigation followed. When the dust settled, Kilby was credited with integrating components on a single chip, and for a working demonstration. Noyce's contribution was in designing a practical planar circuit and a method for interconnecting its components. Jack Kilby acknowledged Noyce's innovation, saying that Noyce's planar manufacturing process provided a reproducibility that assured more consistency than any previous manufacturing process.

The integrated circuits that Kilby and Noyce created laid the foundation for great industries around the globe. It is at the heart of all electronic equipment today, from the pocket calculator to personal computers to the Space Shuttle. Like the invention of the telephone, it has been an innovation that revolutionized society.

How small and how powerful can integrated circuit chips become? Already, workable transistors only one atom thick have been demonstrated in the laboratory, and in early 1992, scientists at the Hughes Research Laboratories in Malibu, California revealed that they had designed and fabricated a transistor capable of operating at the highest frequency ever reported, 300 billion cycles per second.

Hughes Aircraft High-Density Multi-chip Interconnect is next generation of hybrid micro-circuits.

The future of the integrated circuit seems limitless. George Heilmeier, who runs Bellcore, the present incarnation of the old Bell Labs, notes that for the first 25 years after the integrated circuit was invented, the number of transistors per chip doubled every other year, while the cost was cut in half. Had auto makers done as well, by 1991 a Rolls-Royce would cost three dollars and get 100 million miles to the gallon.

Texas Instruments' first hand-held electronic calculator. It was invented by Jack Kilby, Jerry Merryman, and James Van Tassel in 1967.

Chapter 2
The Computer Revolution

Development of computers, and ultimately the application of integrated circuits into microprocessors, set the stage for the Information Age.

S ome time in prehistory, people discovered that information could be preserved and stored outside human memory by painting on cave walls or on animal skins. In time, crude paintings and writings gave birth to alphabets, and animal skins were cured and smoothed to become parchment scrolls. And when the techniques of using the pith and fibers of water plants to make a smooth writing surface was discovered in Egypt, the storage medium advanced from parchment to papyrus.

Humans also needed to keep track of things and to calculate, leading to the development of calendars and number systems that were in use in Egypt as early as 3500 B.C., and in Mesopotamia, where philosophers of the time were solving quadratic equations by 2000 B.C. Mathematics and astronomy flourished in the several centuries before Christ, and as astronomical and mathematical precepts and theorems were developed, they were recorded on papyrus scrolls, which could be indexed and stored in a library or conveyed from one kingdom to another, spreading the information.

The Academy and Lyceum in Athens were great centers of learning dating from the time of Plato and Aristotle, but were closed by the Emperor Justinian in 29. Alexander the Great founded the city of Alexandria on the Nile Delta in 332 B.C. Over time, papyrus scrolls kept in the library there became a major repository of human knowledge. But the library and its 100,000 scrolls was burned when Islamic armies destroyed the city in 640 A.D. The loss of the centers in Athens and Alexandria stalled the march of knowledge in Europe for several centuries.

During the Dark Ages, the creation and diffusion of information were inhibited in Europe, although the thirst for knowledge remained alive. Meanwhile, in China, often ahead of Europe by centuries, advances were kept inside the remote empire. Although the abacus was developed in China around 300 A.D., not until two centuries later was it used in Europe.

In the lands between China and Europe, Islamic culture flourished in the period between the fall of Rome and the beginning of the Renaissance. It became the most advanced culture in the world and spread across northern Africa and onto the Iberian peninsula. Consider the large number of scientific terms of Arabic origin, such as algebra and azimuth.

It was in China and the Arab world that the technique of making paper moved forward. The technique did not reach Europe until during the twelfth century. Before that, information was painstakingly printed on parchment by monks bent over their work with pen and ink in hand. Information recording in the Middle Ages was limited by the speed with which people could write characters or numbers. Perhaps even more importantly, information was spread

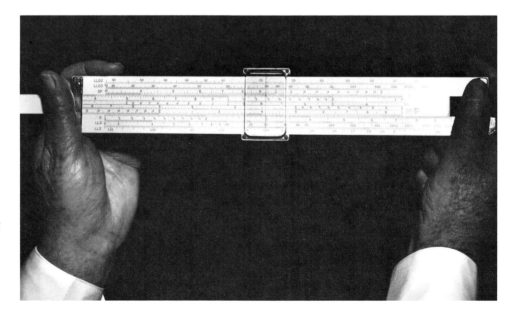

K and E slide rule. This was the standard tool for mathematicians and engineers before computers and calculators.

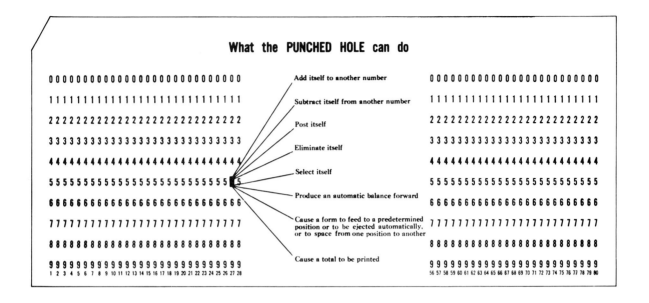

abroad very slowly. It was diffused at the fastest speed attainable by people traveling on foot, by horseback, and in sailboats.

As the need to communicate among cultures increased, people began to search for and develop faster ways of processing and delivering information and to seek better ways to preserve data. The seventeenth, eighteenth, and nineteenth centuries were marked by major advances in these areas. John Napier's discovery of the nature of logarithms eased the tasks of calculating with large numbers, and in 1621 a simple slide rule was invented based on the rules of logarithms. Pascal's adding and subtracting machine followed in 1645, and Leibnitz's calculating machine was invented in 1671; it could multiply and divide.

Charles Babbage designed his Analytical Engine in 1832. It was supposed to be a calculating machine whose operations were controlled by external instructions. Babbage's concept was what today would be called a general-purpose computer, performing operations from a program of instructions set out on perforated cards. Babbage also expected his engine to be equipped with a memory where information would be stored. The device was not built; the technology of the time was just not up to it.

Computer systems used these 80-column punched cards for many years.

Other calculating devices became feasible as technology advanced. The comptometer came along in 1885, developed by Dorr Felt. Its operator used keys to perform addition and subtraction, and a printer was built into the device to preserve the results. William S. Burroughs patented his adding machine in 1888, bringing new calculating power to the office. The comptometer and adding machine were in widespread use in business by the time of the 1890 census, when the need for high-powered calculating machinery became obvious.

Counting Heads in 1890

It is an understatement to say that the population of the United States grew rapidly from the end of the Civil War to the start of the twentieth century. In those 35 years, population growth was explosive and the people had to be counted because of the requirements of the Constitution. Article 1, Section 2 of the Constitution mandated a nationwide census. A major reason for the census: to provide a basis for apportionment of the House of Representatives among the states. The decennial count also was a handy method for gathering information about the inhabitants of the nation. The first census in 1790 counted 3,929,214 persons. By the 1880 census, the population had expanded more than twelvefold and stood at 50,155,783, a 26 percent increase over the 1870 count.

Pioneer computer developer Howard H. Aiken.

For the centennial anniversary count of 1890, members of Congress and state governments nationwide were keen on collecting new social and business statistics. Data on literacy, age, sex, marital status, place of birth, education, and occupation were all potentially important. But collecting and organizing the data and then processing it were monumental tasks.

To those organizing the 1890 census, processing the masses of census information in so many different ways

MARK I

The first digital computer, a machine that performs a controlled sequence of mathematical operations, was the Aiken-IBM Automatic Sequence Controlled Calculator known as Mark I. Designed and constructed between 1939 and 1944, this computer came into being through the cooperation of Professor Howard Aiken of Harvard University and International Business Machines Corp.

Mark I was in operation at the Harvard Computation Laboratory until July 1959 and was disassembled in 1962. Representative parts are reassembled here. One-half of the machine is exhibited at the Harvard Computation Laboratory.

Howard Aiken's Mark I computer, festooned with circuits, as displayed at the Smithsonian in Washington, DC.

seemed insurmountable. However, Herman Hollerith, an engineer from the Massachusetts Institute of Technology, had worked on tabulating the 1880 census and in the intervening decade had developed some ideas for dealing with the problems. Hollerith devised a tabulation system based on punch cards, and he created the machines that enabled operators to punch, tabulate, and sort the cards. Holes were punched in locations that corresponded to an attribute of data. When the punched cards were arranged properly on the tabulating machine, the operator lowered an array of pins, which projected through the holes in the cards and descended into cups of mercury below, closing electrical circuits. An electrical impulse then ran down a wire connecting the circuit to a counter.

The Hollerith process greatly speeded up the counting process. The tabulation took only three years instead of seven. The new device immediately found applications in business and industry, and was soon adopted by census officials around the world. In 1896, Hollerith

*Pioneer
computer
developer J.
Presper Eckert.*

founded the Tabulating Machine Company, which later evolved into International Business Machines.

In 1890, Hollerith's tabulating machine was the most modern device in the world for tabulating and manipulating information. His concept of processing data by holes punched in cards endured for nearly a century, meeting the demands of the industrialized age to use and process more information quickly.

From 1890 through the 1930s, inventors continued to improve ways for calculating numbers. Then, a few years before World War II, Howard Aiken at Harvard University and John Atanasoff at Iowa State independently conceived of the idea of true computers, as opposed to calculating and tabulating machines. Their machines would store, manipulate, and transfer data in large quantities. Atanasoff's machine, which used vacuum tubes and binary numbers, solved large integrated systems of linear equations. The machine read data that was stored on cards by shooting a spark at them. The problem was that some equations required millions of sparks, and a spark misfire occurred about once in every 100,000 firings. This created unacceptable errors. Atanasoff was called into war service before he could perfect the computer.

Meanwhile, Howard Aiken was working on his Mark I, using electromagnetic relays to perform mathematical calculations containing decimal numbers. Harvard University and IBM supported the work, and after five years of development, the Mark I finally operated successfully in 1944. It was big: 50 feet long and five tons in weight. Its more than 765,000 parts were connected by 530 miles of wire.

It was followed by the ENIAC, or Electronic Numerical Integrator and Calculator, developed at the University of Pennsylvania by J. Presper Eckert and John Mauchly of the Moore School of Electrical Engineering. In part, the machine was devised to hasten the speed at

The ENIAC was one of the first machines to be called a computer. It was large as a railroad freight car and weighed 30,000 pounds.

which Army artillery crews could calculate the precise angle for firing their guns in combat.

But the war ended before ENIAC could be put to work to solve the firing tables for the military. Still, it began doing other calculations in 1946. It was boxcar-size (3,000 cu. ft.), weighed 30 tons, and

Intel Computer developer Ted Hoff, who led the team that developed the first off-the-shelf processor.

contained 18,000 vacuum tubes. IBM and Sperry-Rand soon followed with huge vacuum-tube computers of their own design. The IBM 701, introduced in 1952, could execute 17,000 instructions per second, far faster than any electromechanical device.

All the computers of the immediate postwar era were subject to the vagaries of vacuum tubes, which were fragile, short-lived, and prone to overheating. The massive computers were tended by dozens of people, checking tubes and connections and handling millions of punched cards, and seemed suited mainly for huge enterprises. Only the largest military, scientific, and corporate institutions could afford to own and operate them. Most engineers still relied on slide rules to process their mathematical computations.

Making Slide Rules Extinct

As recently as the 1970s, engineers could be identified by the slide rules they carried. The slide rule was really a portable calculator whose operation was based on the concept of logarithms. But the slide rule fell out of use because it was surpassed by a new invention, the hand-held calculator. Jack Kilby and his team at Texas Instruments perfected the calculator in 1967, using integrated circuits. This put extraordinary computing power and speed into the hands of every engineer and student.

Meanwhile, in the area south of San Francisco around Stanford University and Santa Clara (soon to be nicknamed Silicon Valley), Robert Noyce left Fairchild. With Gordon Moore and Andrew Grove, he founded Intel in 1968 to begin producing semiconductor chips. A year later, the company was approached by Busicom, a Japanese company that built advanced calculators, to design a group of chips that Busicom would use in programmable calculators. The task was given to Marcian E. (Ted) Hoff, and a small team of other engineers.

Busicom envisioned using 12 separate chips in a concept that Hoff and his colleagues thought was too complex and expensive. So Ted Hoff came up with a new approach. The Intel team would make a general-purpose computer that was programmed to operate as a calculator.

Familiar Words, New Meanings
Widespread use of computers has given new meanings to familiar words.

Word	Old Meaning	New Meaning **
Address	Where one lives, or gets the mail	Location of information in computer memory
Backup	A vehicle in reverse gear	Saving data files
Basic	Simple	Computer programming language
Bit	Steel piece in a horse's mouth	Binary digit
Boot	One of a pair worn on the feet	Starting the computer
Bug	An insect	Error in a computer program
Byte	To rip a piece from an apple with one's teeth	Eight bits of data
Chip	Residue from chopping wood	Integrated circuit
Crash	Awful mishap with an airplane	Computer becomes inoperable
Disk	Part of the spinal column	Magnetic storage medium
Floppy	What a rabbit's ears do	A type of disk
Hardware	Tools bought at a store of the same name	Physical equipment in a computer system
Icon	Object of devotion	Image, figure, or representation of a process
Mouse	Small rodent	Hand-held control device
Monitor	Armored ship that fought the Merrimac, or someone who oversees a classroom	Video display device
Virus	Something that causes a disease	Program created with malicious intent to damage someone's software
Window	An opening to the outside	Portion of a monitor screen
Worm	Wiggly animal used for fish bait	Nasty computer program that reproduces itself over and over, causing a system to crash

** From PC-Glossary software, Copyright © 1992 Disston Ridge, Inc. Used with permission.

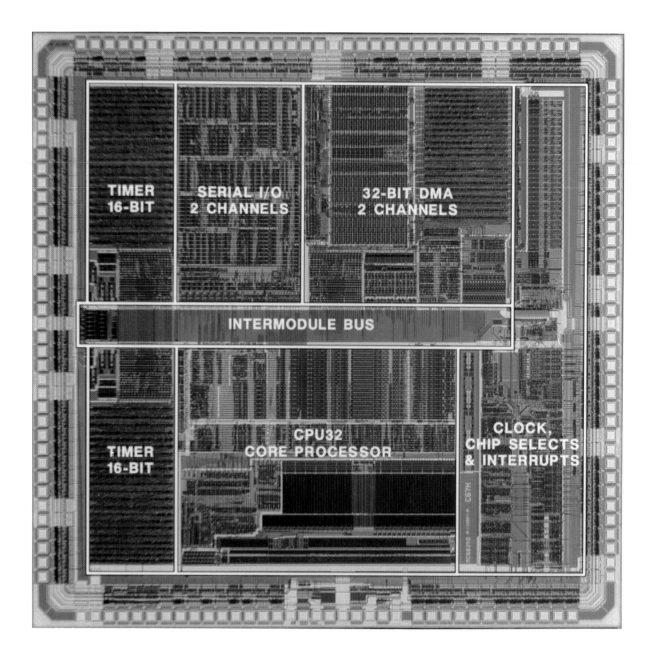

Motorola, Inc. produces this high-powered integrated processor with data manipulation ability.

Hoff and his team went to work to create an off-the-shelf processor that would handle many functions. The package included components that would become familiar to the public in the 1980s and 1990s. At the heart of their design was a chip that acted as a central processing unit, or CPU. It controlled a random access memory (RAM) for processing data, a read-only memory (ROM) for custom applications, and a shift register chip.

The concept was simple, but executing it required ingenuity and hard work. The achievement took nine months. The team assembled 2,300 transistors onto a chip measuring only 3.1 mm (1/8 inch) wide by 4.2 mm (1/6 inch) long. It could perform 60,000 operations per second, acting on 4-bit instructions. The "4-bit" label meant that each instruction contained four bits of code, the combination of 1s and 0s used in digital computers.

The Intel 4004 had as much computing power as the ENIAC of 25 years earlier but was many thousands of times smaller, lighter, cooler, and faster. Best of all, it was easily reprogrammable. Busicom, the Japanese calculator manufacturer, was pleased with the result and eventually sold about 100,000 calculators using the 4004. In fact, Busicom owned the rights to the 4004, but Intel persuaded them to accept repayment of their $60,000 fee in exchange for surrendering exclusive rights. Intel began selling the 4004 to system designers nationwide.

The 4004 was formally introduced at the end of 1971 at a price of $200, and applications proliferated as designers took advantage of

GRiD Systems PalmPad, a rugged, battery-powered computer with a backlit display and solid-state memory.

Stephen Wozniak and Steven Jobs, founders of Apple Computer, revolutionized the personal computer market. By making a smaller, more affordable computer, Apple became a market leader.

what Intel called "affordable computing power." Among the 4004's first applications was its use by TRW, Inc. in the Pioneer 10 interplanetary spacecraft, which blasted off on March 2, 1972 to explore the planet Jupiter. On its 20th anniversary, Pioneer 10, with Intel 4004 microprocessors aboard and its sensors still functioning, had explored Jupiter and sent back pictures and other information about the giant planet. It was eight billion kilometers (five billion miles) from Earth and bound for the stars.

Not to be outdone, Texas Instruments introduced a calculator on a chip in 1971, a breakthrough invented by Michael J. Cochran and Gary Boone. It packed the equivalent of more than 6,000 transistors on a tiny piece of silicon about 6.3 mm (1/4 inch) square. With the new chip, TI could produce electronic calculators at lower prices. The result was a market explosion that heralded the age of the low-cost consumer hand-held calculator. By December 1975, more than 100 million hand-held calculators were in use. Texas Instruments presented its first one (vintage 1967) to the Smithsonian Institution.

Less than a year after the 4004 appeared, Intel introduced the 8008, a microprocessor that could process eight bits of information at a time. Intel's and other microprocessors made smaller computers possible. TI's general-purpose computer on a chip was introduced in 1974; it contained the equivalent of 8,000 transistors in a chip 5 mm (1/5 inch) square. Intel's 8080 chip succeeded the 8008 and became commercially available in 1974. The 8080 cost $360, could execute 290,000 operations per second, and was even faster and more powerful than its predecessor. It quickly became an industry standard, but others such as Motorola developed ever more powerful microprocessors of their own, creating robust competition.

Private individuals were drawn to the new technology and began to put together their own computers. Among the computer enthusiasts were Stephen G. Wozniak (age 26 in 1976), who worked at Hewlett-Packard, and Steven P. Jobs (21 in 1976), who was at Atari. Fourteen years later, Wozniak reminisced about those days:

> I had wanted a computer my whole life; that was the big thing in my life. All of a sudden I realized that microprocessors were cheap enough to let me build one myself. Steve [Jobs] went a little further. Steve saw it as a product that you could actually deliver and sell, and someone else could use.

They sold Jobs's Volkswagen van and Wozniak's Hewlett-Packard programmable calculator to raise $1,350 to get started, and formed the Apple Computer Company on April Fool's Day in 1976. Their first

Electro-mechanical prototype of the mouse, an ingenious solution to a technological challenge.

William Gates, CEO, Chairman and pioneering co-founder of Microsoft Corporation, one of the largest developers of operating systems.

customer was the local Byte Shop computer store, which bought 50 of the Apple I computer boards. The new company achieved early success, especially with hobbyists, and the market broadened when Apple brought out the Apple II in 1978.

A year earlier, across the country at Harvard University, an undergraduate, William H. Gates of Seattle, was working with Paul Allen, a high-school friend, to develop a BASIC programming language for a new Altair computer using the Intel 8080. The project was successful, and Gates and Allen formed a company called Microsoft in 1975 to create and sell software for the blossoming microcomputer field.

The success of the new company's work in software languages, operating systems, and applications meant that it became a de facto standard for microcomputers. The result was a burgeoning of applications using MS-DOS (Microsoft Disk Operating System), as software developers created handy packages based on the system, and customers by the millions began to buy and use them.

The fact that MS-DOS was an accepted standard was a major factor in IBM's selecting the operating system when it began to develop a personal computer. IBM's decision to use an open operating system diverged from past industry practice that had essentially tied customers to a computer company's own system. By using the Intel 16-bit 8088 microprocessor and the MS-DOS operating system, IBM opened the gates to the personal computer rush. Under leadership of Don Estridge, IBM unveiled its first personal computer on August 12, 1981. The price: $1,565. IBM noted that the PC gave users the same power as its mainframe computers had two decades previously.

IBM's success with the PC touched off an explosion in the marketplace. Software developers began writing programs that were compatible with the MS-DOS operating system; hardware manufacturers began producing clones of the IBM PC; and users nationwide

began demanding their own PCs at the office and home. Apple calculated that by the end of 1982, a year after the IBM PC was introduced, more than 100 companies were manufacturing personal computers. By late 1991, according to a *New York Times* estimate, more than 70 million IBM and IBM-compatible personal computers and more than seven million Apple-made computers were in operation worldwide. *Business Week* magazine estimated that personal computer sales in the U.S. leaped from $3.9 billion in 1981 to $37.1 billion in 1991.

An Air Force officer prepares the tasking order for the next day's missions on a laptop computer during the Gulf War.

In the decade after the IBM PC's introduction in 1981, fortunes were made and lost as the global market for personal computers blossomed. Bill Gates of Microsoft became a billionaire in 1987 and was worth more than $6 billion on paper in early 1992, when the stock market valued his company at $22 billion — surpassing General Motors in Wall Street's estimation. Apple prospered too, although to a lesser degree than Microsoft.

Personal computer sales soared as their prices dropped and capabilities improved.

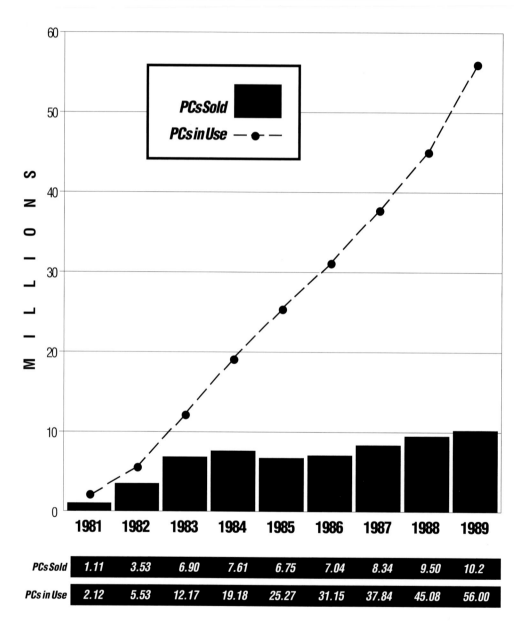

Personal Computer Sales
Domestic United States

	1981	1982	1983	1984	1985	1986	1987	1988	1989
PCs Sold	1.11	3.53	6.90	7.61	6.75	7.04	8.34	9.50	10.2
PCs in Use	2.12	5.53	12.17	19.18	25.27	31.15	37.84	45.08	56.00

As prices dropped and capabilities improved, the number of computer users burgeoned, further fueling competition. Users demanded more portability, larger memory, faster computing power, more applications, and better color displays, and the industry responded.

In the mid-1980s, desktop computers were followed by laptops. At their introduction, laptop computers were portable machines weighing from 10 to more than 20 pounds. The heavier ones were nicknamed "luggables." Over time, they shrank to four to six pounds and fit into a briefcase. In the 1990s, the smaller notebook computers arrived followed by palmtop and coat pocket computers. Personal computers were beginning to be transformed into information appliances.

The computer revolution of the 1980s and early 1990s happened so rapidly and was so widespread that one can lose sight of other technologies that were appearing. When used together with microprocessors, these technologies created the wonders that in 1992 are near-commonplace: communications satellites, lasers, fiber optics, and radar and thermal imaging.

The original synchronous communications satellite, on the Eiffel Tower with Thomas Hudspeth, designer of spacecraft's electronics, and Dr. Harold A. Rosen, leader of the group that conceived and hand-built the satellite.

Chapter 3
A World Without Boundaries

**The wonder of communications satellites
has opened our skies. Images, voices, and
data leap over artificial boundaries, bounced
from Earth to space and back again.**

I n June 1989, the world watched horrified as tanks of the People's
Liberation Army rolled into Beijing's Tienanmen Square, crush-
ing young Chinese demonstrating for freedom. Five months later,
in November 1989, the world exulted when the Berlin Wall crumbled
and fell. In the succeeding two years, Soviet-style Communism collapsed
before the eyes of the world. Mikhail Gorbachev's resignation at Christmas 1991 marked the end of the Soviet Union, and was seen by
watching billions everywhere.

When the protests came in China and freedom surged across
Eastern Europe and the Soviet Union, facsimile messages about the
momentous events flashed into and out of the beleaguered countries,
defying censorship and restrictions. Freedom's race through Eastern
Europe was stimulated and reinforced by images of events transmitted worldwide as they happened.

We now live in a world almost without boundaries. It is a world
in which satellites girdling the Earth are tirelessly receiving and
sending information that may be the television image of a crisis,
streams of data or thousands of simultaneous telephone voice conversations.

Concept of Geosynchronous

Arthur C. Clarke first brought the possibility of communications
satellites to public notice. The time was immediately after the end of
World War II. While still serving in the Royal Air Force, Clarke wrote an

article titled "Extra-Terrestrial Relays" for the October 1945 issue of the journal *Wireless World*. Clarke explained how an object launched from Earth and placed in an orbit 36,000 kilometers (22,300 miles) above the globe would have the same angular velocity as the planet itself. That is, its speed in orbit, 3,075 meters per second (10,090 feet/second), would be sufficiently faster than the Earth's rotation to enable the satellite to hover over a single point on the Earth below. Clarke described this satellite as geosynchronous.

Satellites close to the Earth (in low-Earth orbit) move very quickly, so that the earliest astronauts such as Yuri Gagarin and John Glenn completed an orbit in about 90 minutes. Satellites farther out from the Earth move more slowly. The Moon, for instance, takes about 28 days to make an orbit. Somewhere between those extremes is where the geosynchronous satellite voyages. Its orbital period is 24 hours. Thus it appears to be fixed relative to points on the Earth below.

Clarke pointed out that three such geosynchronous satellites placed in orbit above the equator and spaced 120 degrees apart would provide full global communications coverage to the world's population.

Like most visionaries, Clarke was ahead of his time, and ahead of the technologies needed to turn his ideas into reality. When his article appeared in 1945, the art of rocketry was still crude, and radio communication depended on large, unreliable vacuum tubes. Had a radio communications satellite even been designed and constructed, its weight would have been too great for existing rockets to launch it into the required orbit.

Fortunately, both rocketry and electronics technologies advanced dramatically in the 15 years after Clarke's article. The invention of the transistor at Bell Labs in 1947 freed electronics designers from the tyranny of the vacuum tube. Still, the acceptance of rocketry programs in the West was slow. The newly appointed Royal Astronomer said in 1956: "Space travel is utter bilge."

Rocketry activities in the U.S. accelerated after the shock of the Soviet Union's launch of Sputnik on October 4, 1957. This turned the world upside down and started the space race, infusing U.S. space activities with new urgency. Americans looked up in the sky and saw that small metal ball, made in the U.S.S.R. They heard its beep as it passed overhead in orbit. Suddenly the United States felt truly vulnerable.

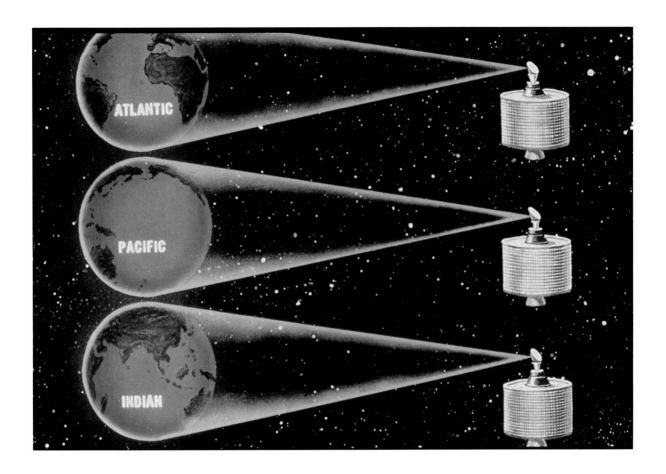

Sputnik abruptly changed the rules of the game of diplomacy, national security, and communications. Soviet manned bombers were already considered a threat to America's security, and now the satellite shook U.S. confidence in its technical superiority. It triggered a massive scientific and engineering effort to regain the high ground. This time, the high ground was in space.

Communications and aerospace companies entered into fierce competition to work with NASA and the Department of Defense. While many companies were supported by government funds in their space development work, other companies, such as Hughes Aircraft, were not.

Three satellites placed in geo-synchronous orbit over the world's major oceans create global coverage.

The Hard Road to Communications Satellite Development

Lawrence A. (Pat) Hyland was Chairman and CEO of Hughes Aircraft when Sputnik shook the world. More than 30 years later, in 1989, Mr. Hyland recalled how he first heard about the Hughes communications satellite program. Two Hughes scientists, Drs. Harold Rosen and Donald Williams, came to Hyland's office one day in 1959 to

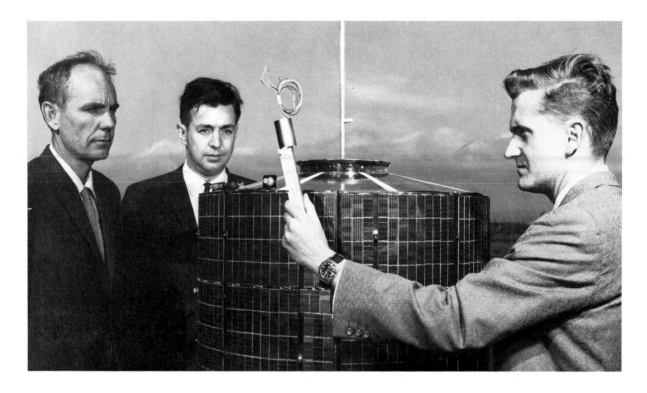

The Syncom Satellite with Hughes Aircraft Company's pioneers on the project: (from left) Thomas Hudspeth, Harold A. Rosen, and Donald D. Williams.

brief him on the project, and to seek corporate funding to continue their work.

Word had already circulated on the inter-office grapevine that the project would cost a lot. Hyland remembered, "I heard that they were coming . . . and that they had some hare-brained scheme for putting up a satellite."

Undaunted, Harold Rosen, the team leader, set up his briefing easel and charts and explained the geosynchronous satellite concept. He described his team's work to Pat Hyland and their need for corporate funds to move the project along.

"They wanted to carry out the theories that had been told by Professor Clarke," Pat Hyland recalled, "and they seemed to be pretty confident about what they were doing." As an experienced engineer, Hyland understood the general idea, but he was skeptical. He focused on their notion of launching the satellite from a spot near the Equator such as Christmas Island in the Pacific Ocean.

Hyland explained to Rosen and Williams that even if their proposition were correct, the diplomatic and logistic obstacles of launching from a British-owned island in the Pacific were formidable.

"As I went on, I kind of liked the sound of my own voice and talking about these immense obstacles, and convinced myself that what I was saying was true," he said. Hyland thought he had also convinced the two young scientists. "After I tired them out, they left a little bit disgruntled, but indicating an understanding that there was some reason in my reply."

Disappointed, Rosen, Williams, and scientist Thomas Hudspeth then each committed $10,000 of their own funds in an effort to get the program started. Learning of this about a month later when the issue was reopened, Hyland was sufficiently impressed that he decided to

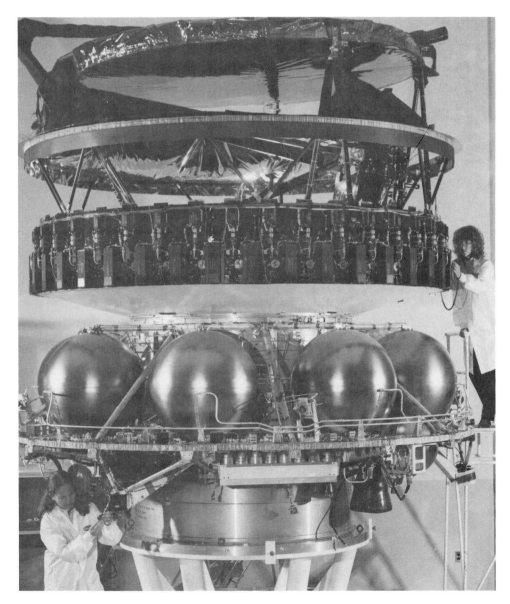

Hughes engineers inspect the inner workings of the Japanese Communications Satellite (JCSAT).

back the project. Pat Hyland realized that "the art had gone beyond me," and the best he could do was to provide company scientists with an environment in which they could work. "They had made the decisions of what they could do," said Hyland, "and I had to back it up or deny it."

The result of this commitment was Syncom, the world's first geosynchronous satellite. Syncom I was launched in February 1963, but exploded when it was hoisted into final orbit.

The next launch attempt came on July 26. Syncom II worked perfectly. The satellite was parked over the Atlantic, and carried out successful communications experiments between the United States and Africa. Syncom demonstrated that satellites were not only commercially efficient but could revolutionize international communications. For the first time, widely separated peoples could be continuously served by communications signals relayed via a satellite constantly pointing at a fixed earth terminal.

Rosen's invention of the spinning synchronous communications satellite was essential to the success of geosynchronous satellites. Donald Williams devised a way to keep the satellite oriented properly and Thomas Hudspeth and John Mendel invented devices that enabled the satellite to receive and retransmit information. From 1963, when the first successful communications satellite was launched, through the mid-1980s, no better alternatives emerged. Even today, derivatives of the first geostationary satellite remain the lowest cost, lowest risk communications satellites aloft.

Irving P. Goldstein, Director General of INTELSAT.

COMSAT Formed

Commercial communications applications of geosynchronous satellites were not long in coming. Congress passed the Communications Act of 1962, and President Kennedy signed it into law on August 31. It authorized creation of a new private company, the Communications Satellite Corporation (COMSAT), designated to represent the U.S. in the global satellite system.

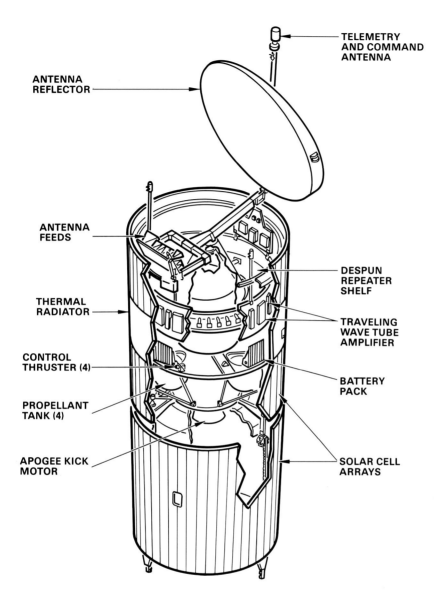

HS 376
SPACECRAFT CONFIGURATION

The company was born on February 1, 1963. COMSAT awarded its first hardware contract on April 16, 1964 to Hughes Aircraft for a satellite to test the feasibility of synchronous orbits for commercial communications satellites. The Early Bird satellite was a resounding success. Launched on April 6, 1965, Early Bird was placed into orbit over the coast of Brazil and a new era of expanded communications began. Its internal design equipped Early Bird to provide 240 telephone circuits, capable of

Japan's first commercial communications satellite, JCSAT, with the highest communications capacity of any domestic satellite outside the U.S.

carrying that many calls simultaneously. This immediately increased telephone capacity over the Atlantic by two-thirds, adding to the capacity that had been available by oceanic cables.

The Beginning of INTELSAT, and More

Soon after COMSAT began operating, the international organization contemplated in the U.S. Communications Act of 1962 came into being as the International Telecommunications Satellite Organization, or INTELSAT. INTELSAT soon placed satellites in orbit over the Atlantic and Pacific. When an INTELSAT III satellite began service over the Indian Ocean on July 1, 1969, three-ocean coverage was established. With that achievement, the most populous regions on the globe were able to send and receive information, including television images, via satellite.

Less than three weeks after INTELSAT III went into service, hundreds of millions of people around the world watched Neil Armstrong and Buzz Aldrin step onto the surface of the Moon on July 20 and sweated out the time until the two explorers rejoined Mike Collins in the Apollo 11 command module for the return voyage to Earth.

Live coverage via satellite has paved the way for other remote events to be viewed from our homes. Members of the global community have had expanding access to scenes of earthquakes, floods, and political and social upheavals as they happen anywhere on the globe. Satellites are also being used to expand education services to remote areas.

Impact of Communications Satellites

Irving Goldstein, former chairman and CEO of COMSAT, became the Director of INTELSAT in early 1992. He noted that communications satellites have overcome five major factors that throughout history imposed limits on mass communications: speed, cost, capacity, access, and Babel.

From the most ancient of times until the early nineteenth century, the rapidity of diffusion of information was limited by the speed of a horse or a sailing ship, and the cost of communication remained high even after the advent of the telegraph and radio. Even after printing presses became capable of churning out thousands of copies of newspapers, magazines, and books, the weight and bulk of a publication imposed limits on its distribution.

Access to important information was traditionally limited to a chosen few, with leaders and governments controlling the flow. The diversity of languages also imposed limits on mass communications. Over the ages, certain languages were used in common to overcome the "Babel effect." The early Roman Catholic church used Latin to transcend local boundaries and dialects, and Latin still is the common language of the Catholic Church. From the Middle Ages to the present, French became the language of diplomacy. In the modern world, English (the language of aviation) is the most widely spoken language in history. One can find English-speakers from the North Pole to the South, and almost everywhere in between. Artificial languages such as Esperanto and various pidgins have been developed in the eternal quest for global communication. None of these languages, however, has come near to solving the Babel problem. But communications satellites have spread the use of English, leading Goldstein in 1988 to forecast the rapid spread of English as a global language.

Irving Goldstein also forecast the arrival of a global currency, perhaps within the next ten years, probably faster than the realization of English as a global language. This global money would not displace local currencies, but would function in addition to them as a means of instantaneous money transfer. A common European currency may become a reality by the mid- to late-1990s. Another prospect that Goldstein forecast was accelerated advancement in the developing world by transmission of energy via satellite. Finally, he said, despite what we forecast, "We will be surprised. No matter how astutely we project, the impact of developments in technology is always surprising."

Arthur C. Clarke too has reflected often on the impact of satellites. Until 1976, making a telephone call to London from his home in Sri Lanka was an exercise in frustration that might last several days. Now, he can get through in the time it takes to punch up the digits. "As a result," says Clarke, "I can live exactly where I please, and have cut my traveling to a fraction of its former value."

Clarke sees a telephone in every village as a realistic and desirable goal by the year 2000. Now that millions of kilometers of expensive copper cable can be replaced by a handful of satellites, transmission is feasible. What is still needed, in Clarke's view, is a simple, rugged handset and solar-powered transceiver plus antenna, which could be produced for tens rather than hundreds of dollars. The solar-powered device would eliminate dependence on batteries, expensive and difficult to obtain in remote areas. As worldwide demand for access to communications continues to build, the prices are certain to drop and capabilities improve.

Arthur C. Clarke lived to see his dream of global communications satellites come true and says, "The long-heralded Global Village is almost upon us, but it will last for only a flickering moment in the history of Mankind. Before we even realize that it has come, it will be superseded by the Global Family," as instant global communications draws populations together.

The setup for Theodore H. Maiman's ruby laser experiment which, on May 15, 1960 at Hughes Research Laboratories in Malibu, California, became the first working laser.

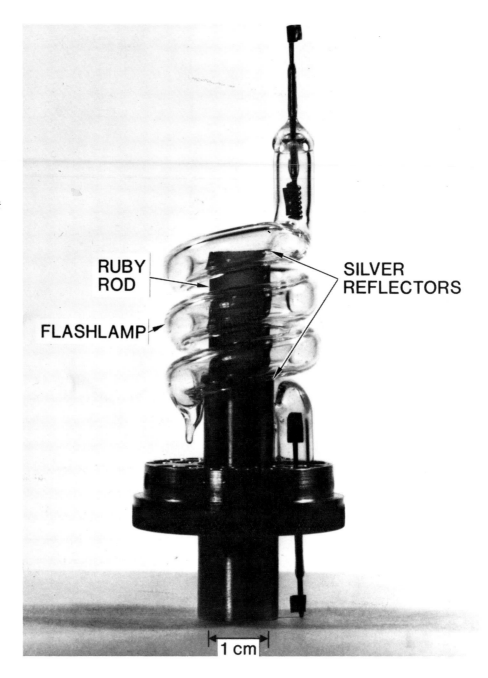

RUBY ROD

FLASHLAMP

SILVER REFLECTORS

1 cm

Chapter 4

Lasers, Lasers, Everywhere

**Lasers have leaped full-force into our lives.
From military weapons and satellite
sensors to operating rooms and the
grocery checkout, lasers are there.**

O n May 15, 1960, the purest beam of light the world had ever seen flashed out of a solid ruby crystal. It illuminated a darkened room at the Hughes Research Laboratories in Malibu, California. Dr. Theodore H. Maiman, with his research assistant Irnee J. D'Haenens and materials researcher Dr. R.C. Pastor, achieved something that had eluded researchers for years.

At that moment more than 30 years ago, the first working ruby laser was born. It was an extraordinary laboratory achievement that excited the scientific community. But the world at large was unaware of its potential. For a time, the laser was a solution in search of a problem. The popular press treated the laser as an exotic flashlight, a death ray, or some other fanciful thing.

Now, the laser's virtues have long since been acknowledged. It is in use everywhere in the developed world, and its applications continue to multiply and expand in such areas as medicine, space research, and communications. Indeed, once working lasers became available, scientists and engineers began applying them to performing jobs no one had visualized.

The word "laser" is an acronym for Light Amplification by Stimulated Emission of Radiation. The light from a laser is intensified and focused much more sharply than natural light or the light shining from an electric bulb. While natural light or light from a bulb is said to be

incoherent, scattering in all directions and covering the entire range of colors, light from a laser is coherent: its beam is of a single color on a narrow wavelength, and the beam is very tightly focused. The light comes from a certain kind of gas or crystal, which, when stimulated by energy such as an electric current, is amplified and releases extra energy as light. The narrow, focused, and powerful beam of a laser can be precisely controlled and finely tuned; it can be used to cut materials, to perform delicate surgery, or to read the bar codes on food packages at a supermarket.

Pervasive as the laser is today, and as simple as its design is, back in the mid-1950s, it was the maser, not the laser, which was considered at the leading edge of research. (Maser stands for Microwave Amplification by Stimulated Emission of Radiation.) The maser was invented in 1954 by Charles H. Townes, James P. Gordon, and H. J. Zeiger. It concentrated microwaves into tightly focused beams of great power that were on the same wavelength. Townes, then on the faculty at Columbia University, and others who followed his line of research, such as Arthur Schawlow of Bell Telephone Laboratories, believed that the techniques that worked to stimulate microwave emissions could be applied to light. In a 1954 book, Townes described lasers.

Indeed, by the late 1950s, most scientists expected that it would be only a matter of time until working lasers were invented. Townes and Schawlow calculated that the laser could be made to work by stimulating monatomic gases such as potassium vapor. They published a paper to that effect in the autumn of 1958, after Bell Labs applied for a patent. Townes continued to pursue the potassium route, while Schawlow at Bell turned to solids, the ruby in particular. But Schawlow and others soon rejected the ruby, based on a belief that its efficiency of fluorescence — 1 percent — was too low to work with the power then

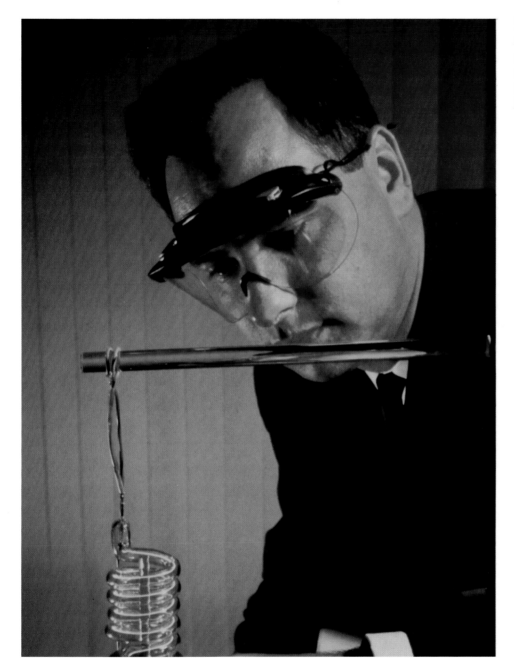

Theodore Maiman keeps a close eye on his ruby laser invention in May 1960.

available. The efficiency figure was based on estimates by Irwin Wieder and his colleagues at the Westinghouse Research Laboratories.

Meanwhile, in the late 1950s, as the Cold War heated up, Hughes Aircraft researchers, along with many other research teams around the world, took up the search for more sensitive microwave amplifiers.

The invention of a gaseous ammonia maser by Townes and his students in 1954 indicated that a molecular-level microwave amplifier was possible. In 1957, Nicholas Bloembergen at Harvard proposed and then demonstrated a more practical solid-state maser amplifier made from cryogenically cooled crystals. Bloembergen's work led to the development of masers made from many different crystals, including synthetic ruby. Scientists at the Hughes Research Laboratories became expert at fabricating ruby masers and subsequently sold several to Cal Tech's Jet Propulsion Laboratory (JPL) for use at the focal points of JPL's giant new deep-space antennas.

Ruby masers have been used for more than three decades to amplify the incredibly small signals received from spacecraft traveling to the edges of the solar system and beyond. Even today, new and improved ruby masers, cooled to operate at temperatures a few degrees above absolute zero (-273 degrees Kelvin) are the most sensitive amplifiers known. Coupled with giant antennas around the world, these devices are used by radio astronomers to detect the debris from the Big Bang at the outermost fringes of the universe.

Maiman and the Ruby

Scientists at Hughes Aircraft in the late 1950s had expanded the use of masers into long-range radar applications and were advancing the art significantly. Like Townes, they and their contemporaries on the East Coast at places like Bell Labs, Columbia, and MIT were convinced that the principles that operated to amplify microwaves (masers) would work with light waves, thus creating lasers.

Theodore Maiman at the Hughes Research Laboratories felt that a solid-state laser might offer advantages over those based on gases. Maiman had joined Hughes in 1956 from Stanford and was familiar with the ruby crystal, having himself created an improved ruby maser in 1959. Maiman had read the literature and, like the others, had stayed away from the ruby. He had tried other crystals, but with unsatisfactory results.

Finally, Maiman decided to check the fluorescent efficiency of the ruby for himself. If he could understand why the efficiency was low, he would use the knowledge to search for a better material. The result was astonishing. Maiman's measurements showed an efficiency of 70 percent. Although the earlier estimate of 1 percent efficiency had led physicists to reject the ruby, Maiman showed it was quite suitable. The discovery was a breakthrough. Dr. George F. Smith, a contemporary and in later years the director of the Hughes Research Laboratories, wrote at the time that "the race was on!"

But the race was not run as quickly as might be expected. Townes and Schawlow had also been examining the pink ruby's possibilities, and they produced an influential paper saying it was unworkable. Management at Hughes had backed Maiman with modest research funds, but now his bosses asked Maiman to abandon the project. He argued for continuation. His analysis and experience with the ruby maser convinced his superiors to continue backing him. Only Maiman's dogged persistence kept this project alive.

Maiman worked on his pink ruby laser model through the winter of 1959-60. To try to raise the energy levels of his laser, he evaluated the intense light sources of the day. Steady high-intensity lights were unsatisfactory, so Maiman selected the GE FT-506 xenon flash lamp.

On May 15, 1960, Maiman obtained his first laser action in a ruby crystal mounted inside the helix of the GE flash lamp. The flash of pure laser light ushered in a new era in technology.

Maiman wrote an article describing the first successful laser experiment and submitted it to the journal *Physical Review Letters*. He titled the article "Optical Maser Action in Ruby." The editors rejected the piece, following a new policy that since the maser field was so mature, only maser articles that contained significant contributions to basic physics would merit speedy publication. They clearly had overlooked the significance of Maiman's paper. Finally, the British journal *Nature* accepted the article and published it in August 1960.

A month earlier, on July 7, Hughes Aircraft Co. had held a press conference at the Delmonico Hotel in New York to proclaim the news and to establish primacy in the field. Maiman showed reporters a ruby laser and explained its broad potential for industrial, chemical, and medical uses.

Research Explosion

News of Maiman's achievement spurred others. Bell Labs scientists replicated Maiman's work with the pink ruby in August. A group at IBM's Watson Research Center succeeded with a uranium laser at Thanksgiving. Schawlow exhibited a laser working with a dark ruby. Another group at Bell Labs (Ali Javan, William R. Bennett, Jr., and Donald R. Herriott) succeeded with a helium-neon laser on December 13. As Joan L. Bromberg noted in her landmark book *The Laser in*

America, 1950-1970, the collective successes added up to a spectacular demonstration that the idea of the laser was sound.

The outburst in research was accompanied by searches for applications. Because the breakthroughs in laser research cut across physics, electrical engineering, and optics, people from a variety of disciplines got involved. Research funds were forthcoming from government and venture capitalists.

Attendance at professional meetings expanded in the early 1960s, as did the number of papers published. The number of companies getting into the laser business swelled. Hughes Aircraft was in the vanguard, joined by other major system-integrating companies such as AT&T, IBM, GE, Raytheon, and Martin Marietta. Hosts of smaller companies conducted research or provided components. Among the smaller ones was Theodore Maiman's own firm, Korad. Maiman and several of his colleagues had left Hughes in 1961 and joined Quantatron, funded by Texas Natural Gas. TNG merged with Allied Chemical, and after a short time, Maiman and his group in 1962 became a subsidiary of Union Carbide called Korad.

Laser research appealed to the public, possibly because it became linked to the cold war. Some associated it with potential beam weapons and death rays. The military saw the potential for new weapons and tools, while civilian applications were predicted in medicine, computers, and communications. Each sector egged on the others, creating a circle of interacting interests and setting in motion a wave of excitement.

Two other developments at Hughes in the first two years of the laser age broadened the application potential for the new devices: Q-switching, which created huge increases in laser power, and stimulated Raman scattering. The concept of Q-switching was based on the notion of suddenly switching the reflectivity of the mirrors at each end of the ruby laser from low to higher values. Robert Hellwarth and Fred McClung of Hughes created the first Q-switching system in 1961. Following Hellwarth's theories, McClung and Hellwarth achieved power outputs a hundred times greater than those of other ruby lasers.

By using a second ruby laser as an amplifier, a group of Hughes scientists found new narrow-beam radiations in the infrared range. Very soon they realized they had discovered a new phenomenon, which they christened "stimulated Raman scattering." This provided a good way to produce new laser colors, and the more colors, the more possible uses. Different colors have different wavelengths, and thus different uses. For example, biologists wanted lasers with short wavelengths (violet), but the Navy wanted blue-green for propagation through water. Both developments opened new vistas for applications.

Laser research continued to accelerate. Basic work at Bell Labs led to successful gas lasers. William K. Bennett, C.K.N. Patel, and others created helium-neon and neon-oxygen lasers and soon expanded into argon, xenon, and krypton gas lasers. Semiconductor lasers soon followed. In the fall of 1962, scientists at General Electric, IBM, and the Lincoln Laboratories at MIT all demonstrated working semiconductor lasers, which had application in optical communications and optical

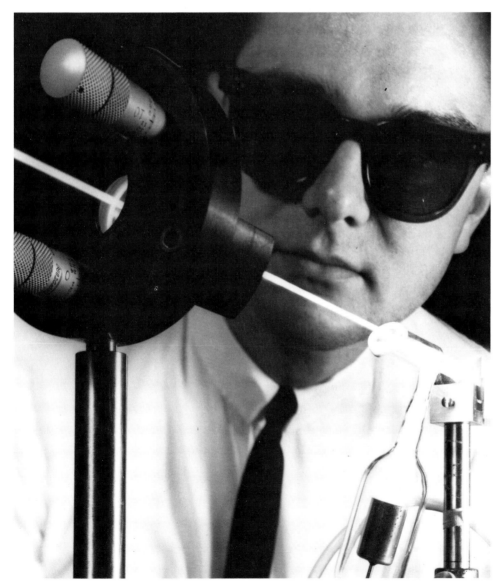

William B. Bridges with an example of the pencil-slim beam of light emitted from a gas laser. His gas laser discoveries expanded the range of lasers and stimulated hundreds of applications.

computing because of their higher frequencies. These frequencies had greater capacity for carrying information. Higher frequencies mean more pulses, hence more bits of information, per second.

The argon-ion laser represented another major step forward. It was the fruit of original work by William B. Bridges at Hughes Aircraft, who demonstrated the first such successful laser on February 14, 1964. By May, Bridges was operating pulsed and continuous-wave lasers using neon, krypton, and xenon in addition to argon. This work broadened the range of visible lasers from only red emitters to a large array of dazzling colors, ranging from yellow to brilliant greens and blues. Blue-green, for example, had potential in underwater communications; low-powered red was found suitable for bar-code scanning; and red laser light created by the argon-ion laser improved retinal surgery.

In one broad stroke, Bridges broadened the laser range to several hundred potential devices, many of which were later commercialized. Malcolm R. Currie, then associate director of Hughes Research Laboratories (and chairman and CEO of Hughes Aircraft, from 1987-1992) said the chemical stability of the inert gases used by Bridges made them highly practical, because they did not react with the elements of the laser tube. The entire visible portion of the spectrum was opened to laser action. Currie saw possible applications for underwater probing, in color displays, and in the laboratory as an instrument that offered a new degree of wavelength freedom for research. Bell Labs built upon Bridges's discoveries to create continuous-wave ion lasers. Perkin-Elmer, Martin Marietta, Raytheon, RCA, and others jumped into the ion laser field.

The carbon dioxide laser, the work of C.K.N. Patel of Bell Labs, came along in 1965. It was followed by the neodymium-doped yttrium aluminum garnet (Nd:YAG) laser, shortened to YAG, a Bell Labs device invented by Joseph E. Geusic.

By the mid-1960s, the roster of operating laser types was both long and productive. Technology and product applications were being implemented as quickly as laser types were discovered. Three major areas emerged as the most fruitful: communications, machining, and measurement of lengths.

Early Applications

Because the Department of Defense and the armed services underwrote a large part of the early laser research, military applications of laser technologies were realized well before the civilian applications evolved in the mid-1960s. For instance, the feasibility of laser rangefinders was proved at Hughes Aircraft by the end of January 1961. Within a short time, Hughes was producing laser rangefinders for the armed services.

The devices changed the way the military operated. The old ways of range estimation and measuring depended upon the experience of the person making the estimation and were often inaccurate. With laser rangefinding, the distance to a target could be determined to within a foot, even at ranges greater than 50,000 feet. Having more

Dr. George Smith of Hughes Aircraft, who perfected the laser rangefinder, holds two early lasers. In his left hand, Maiman's first working ruby laser.

Portable laser scanners are designed to meet the needs of retailers such as supermarkets and depart- ment stores, among others. This is one of several hand- held scanners made by Spectra- Physics.

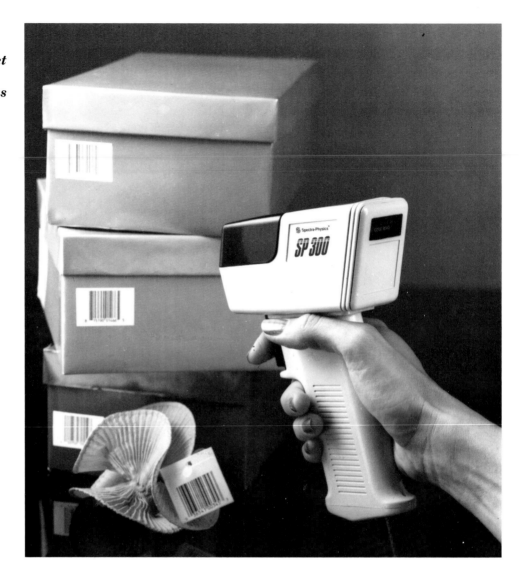

precise range accuracy translated into higher kill probabilities and reduced consumption of tank ammunition and aircraft bombs.

Communications applications for the military and for NASA also came along, but less quickly. Lasers performed splendidly as the light sources for fiber optic communications as that field of technology was developed in the 1970s. Their narrow beam width and short wavelength meant that fiber optic cables conveying high-frequency laser pulses could carry more information than copper wires conveying electromagnetic waves.

Although slow in coming, once commercial laser applications began, they soon multiplied. In December 1961, the first use of a laser to destroy a retinal tumor was achieved in an operation by a team at

Columbia-Presbyterian Hospital in New York using an American Optical Co. (AOC) photocoagulator. The instrument was developed by AOC's Charles J. Koester, along with Dr. Charles J. Campbell, an ophthalmologist and consultant to AOC. The device used a ruby laser as the light source. In materials-processing, communications, and length measurement, laser companies began to develop markets. Lasers were used to machine high-cost materials such as titanium, reducing waste and cost and achieving greater precision. Westinghouse and General Motors were pioneers in this area.

A laser peeler developed by Battelle for the H.J Heinz Company. It uses precisely tuned laser energy to peel thin-skinned foods such as potatoes or fish.

30 years ago, the laser was nothing more than an exotic flashlight.

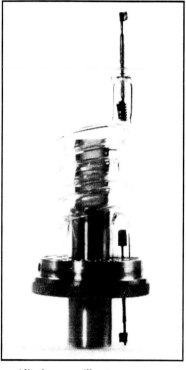

The first working laser- a glass tube coiled around a ruby red rod- was about 3 inches long.

On May 15, 1960, the most intensely pure beam of light the world had ever seen flashed out of a solid ruby crystal.

The laser had been born at Hughes Research Laboratories in Malibu, California.

The Hughes scientists who built that first working laser—and the management behind them—could not have guessed its eventual uses. But now look at what the offspring of their "laboratory curiosity" are doing:

Repairing detached retinas
Reading product codes on groceries
Recording and playing video discs
Cutting fabric for clothes
Drilling holes in metal
Inspecting bottles
Transmitting telephone calls
Surveying roads
Sounding the atmosphere
Dazzling concertgoers
Annealing microcircuits
Welding metal
Characterizing surface roughness
Measuring air pollution
Fingerprinting diamonds
Defining the meter
Slowing atom beams
Cutting cigarettes
Printing computer data
Measuring the earth-moon distance
Cutting airplane parts
Transmitting news wirephotos
Aligning precision machinery
Making three-dimensional pictures
Controlling tunnel machinery
Configuring massive telescopes
Designating military targets
Diagnosing flames
Leveling land
Controlling inventory
Analyzing compounds
Finding impurities

Aligning sawmill cuts
Monitoring polar icecaps
Measuring airplane velocity
Cleaning teeth
Looking for gravitational radiation
Installing acoustical ceilings
Identifying molecules
Aiding robotic vision
Inspecting tires
Positioning medical patients
Probing genetic material
Inspecting textiles
Removing birthmarks
Illuminating fluid flow
Communicating underwater
Enlarging color photographs
Teaching optics
Identifying viruses
Sensing rotation
Peforming microsurgery
Erasing ink
Powering optical computers
Trimming resistors
Altering interconnects
Analyzing materials
Cleaning diamonds
Analyzing auto exhaust
Orienting crystals
Aligning jigs
Ranging targets
Watching continents drift

Sizing dust particles
Cleaning art relics
Tracing air currents
Measuring molecular density
Imploding microfusion pellets
Sensing cloud altitude
Monitoring earthquakes
Gauging fine wines
Testing optical components
Analyzing thin film composition
Drilling holes in diamond dies
Testing relativity
Separating isotopes
Sensing liquid level
Sensing magnetic fields
Programming read-only memories
Counting blood cells
Guiding missiles
Gauging film thickness
Monitoring crystal growth
Aligning large optics
Shaping jewel bearings
Measuring the speed of light
Securing perimeters
Positioning x-y stages
Computing in parallel
Pumping hard-to-pump lasers
Astonishing moviegoers
Creating highly excited atoms
Amplifying images
Cauterizing blood vessels
Diagnosing fusion plasmas
Enhancing chemical reactions
Engraving identification marks
Hardening surfaces
Perforating computer paper
Producing advertisements

Imagine what the next 30 years will bring

Bell Labs, GE, Hughes, Raytheon, and others with much experience in electronics found new ways to use lasers for communications. Nearly three decades later, though, there are no military or NASA laser communications in space. Eventually, some satellite-to-satellite laser communications may be realized, but the experts predict that satellite-to-earth laser communications are unlikely, since laser beams cannot transmit through clouds or heavy fog.

Honors came both early and late to the major players in the development of the laser. Charles Townes shared the 1964 Nobel Prize in physics with two physicists from Moscow, Alexander M. Prokhorov and Nikolai G. Basov, for fundamental work in quantum mechanics leading to maser and laser devices. Theodore Maiman was inducted into the Inventors Hall of Fame in 1984, 24 years after his success with the pink ruby laser.

In 1990, on the laser's thirtieth anniversary, scientists at the Hughes Research Labs where Maiman and his group had operated the first successful laser paused to compile a list of laser applications. They headed the list with this statement: "Thirty years ago the laser was little more than an exotic flashlight." Their list quickly reached 103 applications (see table) and continues to grow each year.

Today, lasers are truly ubiquitous, although not always obvious. Scanners that capture bar-code data are widely used in supermarkets and retail stores. But neither clerks nor customers are aware that a laser is at the heart of the scanner. Federal Express Corp. employees log in more than a million shipments on an average day through the use of hand-held laser scanners and palmtop computers. Laser technology has also advanced surgical and dental techniques, as well as desktop publishing and instructional processes. And in the kitchen of the future, a real time-saver may be the state-of-the-art laser potato peeler perfected by Battelle Institute for the Heinz Corp.

The laser in its many forms has truly become the light fantastic.

Researchers Linn Mollenauer (foreground) and Michael Neubelt of AT&T Bell Laboratories work on lab experiments that measure bit-error rates in transmission of solitons (light pulses that travel distortion-free over long distances). Solitons promise to multiply the capacity of communications circuits many-fold.

Chapter 5
Harnessing Light for Humanity

Light and glass create information highways of near-infinite capacity, another example of science converting sand into technical power.

O ptical communications are as old as signal fires on a hilltop or a lighthouse along a hazardous coast. The signal lamps in Boston's Old North Church sent the message to Paul Revere that the British were coming. But today, optical communications have taken a new direction, as new as the bundles of ultra-pure fibers of glass carrying trillions of bits of data on pulses of laser light.

Sand, transformed into pure strands of silica fiber, has supplanted copper cables as a high-capacity medium for conveying information throughout the countryside and across the oceans. Communications via fiber optic cables are more efficient than radio and carry more information than copper. Unlike the spreading beams from radio antennas, glass fibers contain light energy so that it travels in precise paths over great distances. Because it operates at higher frequencies, the laser light transmitted over fiber optic networks carries many times more information than electromagnetic pulses pushed through comparable copper cables. Higher-frequency light waves are divided into more pulses per second. The pulses, and the gaps between them, create the 1s and 0s of digital information. Thus the higher the frequency, the more information can be carried.

The concept of capacity means a lot in communications. For instance, the human brain's memory capacity seems to be limitless. But the tribal wise ones of ancient times could remember only a limited amount of tales and information to hand down to the young in stories told around a fire. Later in humanity's evolution, the information that could be carried by messengers on foot or horseback was limited by weight. Information sent by ship could be voluminous, but its transmission time was determined by the ship's maximum speed through the water.

Invention of the telegraph increased the speed of transmission of information close to that of electricity coursing through copper wires. Although the telegraph's capacity may have been limited, its higher speed gave it an edge. Telegraph lines soon spread among cities east of the Mississippi in the years before the Civil War. Raising money for the first transatlantic telegraph cable was a feasible venture in the mid-1850s, and the first such cable was laid in 1858 under the leadership of Cyrus W. Field. Soon after the cable was completed, Queen Victoria and President James Buchanan exchanged greetings, but within a few weeks the cable snapped.

The Civil War intervened to delay laying the replacement cable across the Atlantic, but it was finally completed in 1866. It was strung off huge drums mounted on the stern of the ship *Star of the East*. The arm-thick cable of copper wire, wrapped in burlap and gutta-percha (a rubber-like compound today used for insulation and waterproofing), linked England and North America by telegraph. The cable's capacity: one message in one direction at one time. Limited as that capacity might sound today, it opened a period of instant telegraphic communication across the Atlantic. Information in the form of telegraph pulses crossed the water at the speed of electricity flowing through copper, instead of requiring ten days to two weeks aboard ship. Location of the European terminus in Great Britain created a communications hub and gave Britain an information edge over its commercial competitors on the continent.

When Guglielmo Marconi demonstrated the feasibility of wireless telegraphy, or radio, the discovery offered the potential of greater capacity and near-instantaneous speed, as well as freedom from fixed cable sites. His transmitting station at Poldhu in Wales achieved transatlantic wireless telegraphy communication in 1901. Wireless telegraphy systems of Marconi's time were based upon sparking, an application of Heinrich Hertz's discovery of electromagnetic waves in

1886. A spark at the transmitter generated short electromagnetic waves that leaped to and fro, or oscillated. The wavelengths were short enough to be focused by a reasonably sized antenna, but long enough to travel great distances through the air. If sent out with enough force from the transmitter, the waves could be detected by a distant receiver with a similar antenna.

When the characteristics of the frequency spectrum and its behavior were better known, scientists realized that light waves would be able to carry more information than waves at lower frequencies in the spectrum. But in the late nineteenth century, no practical optical communications systems were available.

Communication by light waves has been a goal of scientists at least since Alexander Graham Bell, who invented a Photophone soon after he created the telephone. His Photophone transmitted voice signals along beams of light reflected off mirrors. Bell demonstrated the Photophone in February 1880. His invention failed, however, because the mirrors worked only in crystal-clear atmospheric conditions. Smoke, haze, and the innate properties of air rendered the instrument useless.

Bending Light Around Corners

Ten years before Bell demonstrated his Photophone, a British physicist guided light around a corner. In 1870, John Tyndall shone light into a parabolic stream of water, making it curve with the stream. It worked thanks to a phenomenon called total internal reflection. When light traveling through a dense medium such as water strikes a less-dense medium such as air at a small angle, it is reflected into the denser medium. The boundary between the denser and lighter media acts like a mirror. The index of refraction of a medium is a measure of how light is refracted when passing through it. In this case, the index of refraction of the light-carrying medium (the core) needs to be higher than the outer shell for the phenomenon to work. The refractive index of water is higher than that of air, thus Tyndall could bend light.

The work of Tyndall and Bell stimulated others to attempt practical optical communication. But all were thwarted by the limitations on light. They tried hollow pipes with reflecting surfaces and a host of other ideas. But the light diffused, that is, it spread out, and the signals were lost within a few feet. Dr. David A. Duke of Corning notes that all the elements for a successful optical communications

Dr. Robert Maurer, together with Dr. Donald Keck, is responsible for developing low loss optical fiber in Corning's laboratory in 1970.

system had been identified in the late nineteenth century. But it was not until 1971 that a confluence of technical advances and the inspiration of a team of three scientists at Corning made optical communications commercially feasible.

1971 — The Time is Right

Until the invention of the laser at Hughes Research Laboratories in 1960, light from existing sources that was transmitted through glass tended to scatter. Maiman's ruby laser and the other solid and gaseous lasers that soon followed changed that. The laser's narrow monochromatic beam of coherent light emitted waves in phase with each other. It was an ideal carrier for communication signals.

Glass fibers were seen early as potentially fine conveyors of light signals; suitable optical fibers had existed since the 1950s. The problem was that even with the laser's coherent beam of light, a signal could not be sent very far without a repeater (or amplifier) to boost it back up. Otherwise, too much of the signal would be lost in passage. The optical fibers of the day may have been suitable in theory, but too much of the signal was lost in too short a distance for practical, economically attractive communications.

Two scientists at the Standard Telecommunications Laboratories in Britain, Charles K. Kao and George A. Hockham, speculated that impurities in the fibers available in the mid- to late-1960s scattered the light. This scattering (or attenuation) meant that even over a short distance of less than one kilometer, not enough of a signal remained for practical communications. Remove the impurities, it was thought, and it would be possible to send information over the optic fibers far enough to make the process practical. Kao and Hockham showed that an attenuation of 20 decibels or less per kilometer would permit commercial use of optical fibers for telecommunications. Amplifiers would be needed at frequent intervals to boost the signal back up, but the system would work.

Attenuation in Glass Fibers

In fiber optics, the decibel (dB) is a measure of the fraction of the signal that makes its way through a communication system. The glass fibers of the late 1960s had attenuation of 1,000 decibels per kilometer. That meant that only one part in ten to the hundredth of the signal power would emerge, and this was far too little to be practical.

The British were stumped. If glass could be developed free enough of impurities to have an attenuation level of 20 dB/km or less, a system could be made operative. In such a fiber, 1 percent of the signal would emerge after a kilometer of travel, and that was enough to be amplified and repeated down the line. But such a glass did not exist. The researchers mentioned their interest to a scientist from Corning Glass Works who was visiting as part of a research sabbatical. He carried the word back to Corning, and the project was assigned to a research group headed by Dr. Robert D. Maurer, a physicist who joined Corning in 1952. He worked first on ultrasonics, then joined the fundamental research department to work on light-scattering studies in glass. When the British challenge was handed to him, he was manager of fundamental physics research and had not been involved in fiber optics. When Maurer later reflected on his achievement, he said that not having been involved in fiber optics work probably helped him find a fresh approach, unconstrained by the feeling that the challenge was almost impossible.

Donald B. Keck was relatively new at Corning when Maurer involved him in the fiber optic research. Peter C. Schultz, the third member of the team, was a chemical wiz-

Dr. Donald B. Keck, together with Dr. Robert Maurer, is responsible for inventing low-loss optical fiber in 1970.

ard who created blends of silica for the experimental work.

The conventional wisdom of the scientific community argued for using compound glasses. They were the obvious choice: easy to melt, thus easy to draw into long lengths of fiber. Many kinds were available, so finding something with the right refractive indexes ought to work. But it didn't.

Maurer and his team avoided the obvious — and solved the problem. They used fused silica. It was especially pure, but had disadvantages: a high melting point and the lowest refractive index of all the available glasses. That should have ruled silica out. Remember that the core carrying the light beam should have a higher refractive index than the outside covering, or cladding. How could that apparent obstacle be overcome? The elegant solution devised by Maurer and his team relied on a contrarian approach. They used silica for the entire fiber, but raised the refractive index of the fused silica core by intentionally adding a dopant, or impurity, to it. The material they turned to was titanium dioxide. With the resulting impure core having a higher refractive index than the pure outer shell, they theorized that light focused into the core should stay inside and the system should work. It did.

Maurer, Keck, and Schultz broke the 20 dB/km barrier in the summer of 1970, and reported it at a British telecommunications conference at the end of September. Fiber optic communications was now practical. A new era was born.

Once the Maurer, Keck, Schultz team showed it could be done, others quickly picked up on their discoveries. But the Corning team did not relax. By 1972, they announced a graded-index fiber with a loss of only 4 decibels per kilometer. By 1991, minimum fiber loss in telecommunications cables was reduced to 0.16 dB per kilometer. That is more than a hundred times lower than the 1970 version that broke through the 20 dB barrier. The signal traveled significantly farther before being boosted and sent onward.

Spinning the Glass Web

The marriage of laser light and fiber optics created another explosion of applications, since the light waves could carry so much more information in ever-smaller diameter cables than was previously possible. The laser field had advanced enough by the time of the practical fiber optic cable in 1970 to offer a wide range of choices for suitable laser light sources. Semiconductor diode lasers were the best suited. Light-emitting diodes (LEDs) were another suitable source. As it turned out, both types emitted light at wavelengths between 800 and 900 nanometers (a nanometer is one-billionth of a meter), a range where optical fibers transmitted well. Solving the challenge of turning light signals into electrical signals at the receiving end was fairly straightforward because silicon detectors sensitive at 800 to 900 nanometers were readily available.

Coupled with the information processing and display powers made possible by ever-smaller and more powerful microprocessors, fiber optics applications soon appeared in medical diagnostics and microsurgery. Doctors using micro-video cameras at the end of flexible fiber optic cables were able to see inside the human body and could make earlier and more precise diagnoses than before. In the late 1970s and the 1980s, the endoscope came into being. The instrument went down the esophagus into the stomach. Sigmoidoscopy evolved as physicians were able to view the colon and lower intestine.

Fiber optic cables such as the one shown offer the promise of creating "highways of information."

Cystoscopy allowed doctors to see along the urethra and into the bladder.

Telecommunications applications of fiber optics came about even more quickly. The first fiber phone system was installed in Chicago in 1977. Eleven years later, the first transatlantic fiber optic telephone cable, TAT-8, began operating. With two fibers in each direction, it carries up to forty thousand telephone calls simultaneously. In 1989, transpacific fiber optic cable service was introduced.

Enter the Soliton

The uses of optical fiber in communications and elsewhere became commonplace in the two decades following the breakthrough by Maurer and Keck. By the early 1990s, signals could be sent up to 100 kilometers (62.5 miles) before being boosted electronically and sent onward. Even with the extraordinary progress, transmission through optical fiber still was limited by two key factors, dispersion and signal strength. The pulses of light in fiber actually contain more than one frequency, and the higher-frequency components moved faster than the lower ones. Over a distance, then, the components of the pulse dispersed or broadened, eventually losing their accuracy. As for signal strength, it decayed over distance, a fact that frustrated early researchers.

Researcher Linn Mollenauer of Bell Labs sought a means to create an all-optical transmission system to overcome these limitations. For the solution, Mollenauer turned to solitons, pulses that maintain their shape over long distances. The soliton phenomenon was first described by John S. Russell, a Scottish shipbuilder of the late nineteenth century, who observed a wave that passed through a canal and traveled as far as he could see without losing its shape. In 1973, Akira Kasegawa of Bell Labs was the first to suggest that solitons could exist in optical fibers; he proposed the idea of a soliton-based transmission system. Since the speed of light in fiber depends on its intensity, intensity could be used to cancel the effects of dispersion, creating a soliton.

Mollenauer devised experiments that exploited the characteristics of solitons, using separate pump lasers to give the signals an optical boost as they traveled. In February 1988, Mollenauer and his colleague Kevin Smith sent light pulses through fiber for a distance of 4,000 kilometers (2,500 miles) without electronic regeneration. The same optical signal put in one end came out the other with no loss. Mollenauer and his colleagues later transmitted solitons more than 11,900 kilometers (7,440 miles).

By 1991, a team at Bell Labs headed by Anders Olsson transmitted solitons in an experiment at 32 gigabits (billion bits per second). That was about thirteen times faster than the 2.5 gigabit transmission speeds used by telecommunications companies.

The implications of these advances for high-capacity long-distance communications are enormous. As capacity increases, the variety of services will increase, and the cost of communications will go down. Video phones, high-definition television, and consumer videotext are all examples of the seemingly limitless range of services made possible by advances in fiber optics. As Mollenauer said in 1991, "We've begun to discover that as we add capacity to a system, the world soon finds ways to use it and even demands more."

What of the future? Robert Maurer, who described his development of the first low-loss fibers as an "interesting bit of work," predicted that fiber optics is going to take over almost all of communications. Both he and Mollenauer recognized that widespread applications of all-optical systems await development of optical switches, connectors, couplers, and other ancillary components. Given the surge of research in photonics of the 1980s and 1990s, those days are probably close at hand.

Just as fiber optics discoveries coupled with laser light transformed communications, so did advances in sensors bring great changes that transformed the world of combat.

Massive antenna array of U.S. Army's SCR270 radar was manufact- ured by West- inghouse and installed on Oahu, Hawaii, in 1941. Its operators detected large formations of aircraft approaching on the morning of December 7, but their warning was discounted.

Chapter 6
Seeing in the Dark

**Radar and infrared sensors have
freed humankind from the
blindfolds of night and weather.**

T he technologies discussed in earlier chapters created new ways of processing and conveying information. Each began as a single breakthrough by an individual or small team of scientists. They achieved the feat before contemporaries who were also seeking solutions to the same challenges.

In this chapter, two enabling technologies of a different nature are discussed: radar and infrared, also called thermal imaging. These technologies *collect* information and extend our own senses. It is difficult if not impossible to cite specific individuals or identify single breakthroughs for these technologies as with the earlier technologies, because so many laboratories, companies, and government agencies were involved and these areas developed along so many lines. In these cases, as well as others cited in later chapters, advances in one area of electronics technology led to progress in other areas. With that introduction, let us investigate how radar and infrared devices developed and flourished in the past four decades.

Introduction to Radar

The sailors standing lookout on Christopher Columbus's flagship *Santa Maria* were balanced 53 feet above the Atlantic Ocean. From that lofty perch, they strained their eyes to make the first sighting of what they thought was India, the continent that Columbus expected to reach. When they actually made the first sighting of land in the Bahamas, they were 23 to 27 nautical miles away. They could only see 20.5 miles, but atmospheric refraction carried their vision a bit farther. Beyond that distance, the curvature of the Earth hid everything behind the horizon.

Columbus and his sailors bumped into the New World 500 years ago. They did not know precisely what it was or where it was. For them and other seafarers for centuries before and after, navigation was a less-than-precise art. However, 477 years later, on July 20, 1969, Neil Armstrong and Buzz Aldrin landed on the Moon because they knew exactly where it was, and, moreover, knew exactly where they were landing and what the landing area looked like. They enjoyed such precision because radar had mapped the landing area, guided their voyage from Earth to Moon, and enabled them to make a precise descent to Tranquility Base. When they were ready to leave, radar provided the guidance link between the first two men on the Moon and their astronaut colleague Mike Collins awaiting them in lunar orbit in the command module, the spaceship *Columbia*.

Radar enables people to see far over the horizon and to pierce the darkness. It has also been used to range through the solar system and the galaxy, and to map the planets and find distant stars. Here on Earth, radar is one of the tools that each day permit millions of air travelers to fly through darkness and bad weather and arrive safely at their destinations.

The word "radar" is an acronym derived from World War II usage, and it stands for RAdio Detection And Ranging. Its operation is based on the principle of the echo. George Stimson explains the principle in his book *Introduction to Airborne Radar* by giving three examples: a blind man walking along a sidewalk and tapping his cane to remain a fixed distance from a building wall; a bat sending out shrill beeps to locate and swoop down on insects for a meal; and a supersonic fighter aircraft closing on an enemy plane hidden behind clouds 100 miles away.

The blind man, the bat, and the fighter airplane are all detecting the location of objects and determining their distances by interpreting the echoes the target objects reflect to them. The difference is that the blind man's cane and the bat send out sound waves, while the fighter's radar set emits radio waves. They all work because of reflectivity. That is, most objects reflect waves of the electromagnetic spectrum, whether the waves are sound, radio, or light.

The reflecting phenomenon of radio waves has been known since 1886, when the German scientist Heinrich Hertz demonstrated the principle. Not long afterward, a German patent was issued for a system that used radio echoes for ship navigation. But the knowledge was not put to practical use until the mid-1930s, when early radar systems came into being. Radio had come of age during World War I, thanks to use by the military. In the 1920s, it was exploited for commercial and entertainment uses. In the interwar period, research on radio continued to expand its utility for both military and commercial purposes. Radar had a harder road to travel.

In the early 1930s, work on radar was pursued mainly in Great Britain, Germany, and the United States. Technically, the Germans led, with the British next and the Americans lagging. Germany and Britain both were developing coastal surveillance direction-finding systems to alert their respective territories to impending air attack. Robert Watson-Watt headed the team developing Britain's area-defense radar, a system that was ready by September 1939.

In the United States, the Navy Research Laboratories (NRL) in Washington were pursuing development of high-frequency direction-finding equipment. NRL was (and still is) on the east bank of the Potomac a few miles south of Washington, D.C. It was immediately south and in the traffic pattern of two busy military airfields, the Navy's Anacostia Air Station and the adjacent Bolling Field of the Army Air Corps. In their research, the lab workers at NRL observed and verified the reflection of high-frequency radio waves from passing aircraft. Scientists at Bell Labs observed the same thing at Crawford Hill, New Jersey, in the mid-1930s.

Robert M. Page, one of the Navy scientists on the banks of the Potomac, constructed the first pulse radar set in 1934. A pulsed radar sends out radio waves in pulses of energy. There is a tiny gap between the pulses. The radar's receiver listens for the echoes during the intervals between pulses. Page and his associate Leo C. Young made the first detection of an airplane with the pulse radar set.

Another of the navy researchers was a civilian named Lawrence A. (Pat) Hyland. He had served in the Army Signal Corps in World War I and later spent six years as a navy radioman aboard ships and aircraft. Hyland and his colleagues at NRL foresaw immediate military utility for radar and requested money from the Navy's Bureau of Ships to pursue their discoveries. But Bureau officials did not believe the NRL radar discoveries were very important. They were not forthcoming with funds. However, the authorities did provide enough money for Page to install and test a radar set aboard a destroyer in 1937.

Hyland used his own equipment to advance the research. His achievements included the first successful identification of a dirigible via reflected radio waves. He thought this was so important he sent a report to the Bureau of Ships. But, Hyland recalled, "the chief of that group came down personally and told me to stop; that this had no possible application for military purposes." Hyland resigned from the NRL, applied for a patent on his discoveries, and went into commercial radio research in the mid-1930s. He carved out a distinguished career with Bendix and was hired by Howard Hughes himself in 1954 to manage the Hughes Aircraft Co. After Hughes died in 1976, Hyland became president and then chairman and CEO.

As the decade of the 1930s ended, the U.S. Army Signal Corps was pursuing radar development more vigorously than the Navy. The Army ordered its first long-range search radars in 1940. Westinghouse Electric Corp. was the supplier; it delivered the first six SCR270 sets in 1941. They were shipped to Hawaii that summer and put into operation. By early December 1941, radar operators were trained and using the sets to search north and west of the Hawaiian Islands.

On the morning of December 7, one of the radar operators saw indications on his scope that signified a large formation of airplanes at a distance of 137 miles They were coming from the north and headed toward Oahu. The radar operator telephoned the information to the warning center, but the officer on duty there concluded that the airplanes were probably a group of B-17 bombers arriving from California. Fifty minutes after the radar operator passed on the information, the first Japanese dive bombers attacked Pearl Harbor and other targets on Oahu.

On that fateful Sunday morning, the radar set and the operator had done their jobs; but the system for evaluating their information and turning it into useful intelligence failed. More than a year earlier, radar of similar design was operational along the south coast of England and was used more effectively as part of a defense system, enabling the Royal Air Force to marshal its few Hurricane and Spitfire fighter aircraft against oncoming Luftwaffe fighters and bombers in the Battle of Britain.

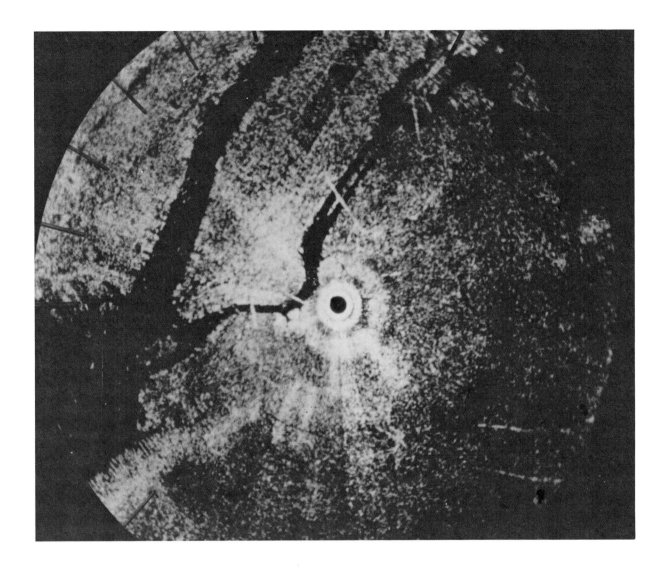

From the benchmark events of the Battle of Britain and Pearl Harbor, radar development surged throughout World War II. "The wizard war," as Winston Churchill called it, marshaled the scientific and technological resources of both sides. In the process, radar matured quickly and became an indispensable element of warfare.

Manhattan Island and vicinity from the radar display aboard a U.S. Air Force B-17 bomber in the late 1940s.

Detecting Moving Targets

Spurred by new advances in technology, the pulses of radar could detect ever-smaller targets and provide reasonably accurate determination of the range over ever-greater distances. Still needed, however, was a way of discriminating moving targets from stationary ones and a means of determining the speed and direction of moving targets. The answer lay in the Doppler effect, named after the German scientist Christian J. Doppler.

To human ears, the Doppler effect is exemplified by the raising or lowering of the pitch of the sound emitted by a moving car horn, train whistle, or jet aircraft. As the source of the sound approaches the listener, its pitch increases. It sounds higher as the sound wave is compressed and its frequency rises. As the source speeds away from the listener, its pitch decreases. It sounds lower as the sound wave is spread out and its frequency declines. This occurs because more wave cycles reach the ear in one second from an approaching vehicle than a stationary one. Say the train whistle blows at 300 cycles per second (300 Hz; Hz is short for Hertz, after Heinrich Hertz) when the train is not moving. If it is speeding toward a person, sounds at perhaps 360 Hz are reaching that listener's ear, and the whistle sounds higher than if it were immobile next to him. After the train passes and is speeding away, sound waves at perhaps only 260 Hz are reaching the listener, so it sounds lower and even mournful.

The technology of advanced radar and the skills of air traffic controllers and aircrews enable air transport operations to continue around the clock and through bad weather.

The same effect applies to radio waves. If the pulses sent out from a radar are radiated back to it as pulses at a different frequency, then the target can be discriminated from the background, and its speed can be determined. Both sides solved this technical challenge in World War II. As radar and other technologies advanced, processors and memory-storage devices were added to the basics.

But in the 1950s, size, power requirements, and displays presented serious challenges. Size was a problem because antennas had to be large to detect and collect weak returning pulses. Antennas needed mechanical linkages to rotate or point them in the desired directions. High power was necessary to push out signals strong enough to hit targets at longer ranges and still create enough reflective pulses to be detected. The magnetron made high-power microwave radars possible. In a radar transmitter, it amplified relatively low-input direct-current voltages into the high radio frequency power levels needed for efficient operation. In everyday life, the magnetron also found a home in millions of kitchens; it is what powers modern microwave ovens.

Displays, or means of interpreting the data that the radar set provided, required constant improvement to increase radar's usefulness. The earliest displays simply showed the outgoing pulse as a sharp upward rise in a glowing horizontal line across a cathode ray tube (or Braun tube, after its German inventor). The returning pulse showed up a bit later on the line as a smaller but discernible rise. The rises in the line caused by the radar pulses were called blips. Range to the target was the only information those displays provided.

Later World War II-era radar displays placed the operator at the center of the display screen, and the antenna swept around continuously. In the display, the radar's sweep was represented by a glowing radius of a circle that rotated in synchronization with the antenna's search. On a radar scope called a Plan Position Indicator (PPI), both range and direction to targets were displayed.

Whatever the display, operator experience played a significant role in the utility of the system. An experienced operator could wring out more information, while a novice was baffled by the primitive radar displays of WW II and the postwar years. Years of peering at dancing lines and pulsating blips took a toll on radar operators. It was no accident that they were nicknamed scope dopes.

Applications Flourished

In the postwar era, as commercial air operations flourished, ground-based radar sets overcame the twin nemeses of safe flying operations, darkness and weather. During the Berlin Airlift of 1948-49, ground-controlled approach (GCA) radar guidance permitted aircraft to land under conditions impossible during World War II. For the GCA approach, ground-based radar at Tempelhof airport in Berlin and Rhein-Main airport at Frankfurt provided information on an aircraft's position in relation to the proper descent path leading to the runway. Observing information displayed on the radar scope, the air traffic controller transmitted instructions to the pilot by radio. In the cockpit of the transport aircraft approaching to land, one of the pilots concentrated on the aircraft instruments and applied corrections, while the other watched out the windshield to spot the runway or its lights through the gloom. The controller continued to transmit instructions until the pilots saw the runway and could make the final corrections for a landing.

The GCA radar sets and the procedures used in the Berlin Airlift soon spread into commercial use at major airports. For example, radar approach and departure procedures were put into effect in the Washington, D.C., air route traffic control center area in 1952. Increased use of radar within the air traffic control system of the United States, and improvements over the years, enabled airlines to fly more frequently at night and in bad weather that earlier would have grounded their aircraft. The surging growth in air travel would not have been possible without reliable modern radar, both on the ground and in the aircraft themselves.

Both in the United States and abroad, many electronics companies contributed to radar development and the production of radar equipment for both commercial and military uses. Major British radar manufacturers included Cossor, Ferranti, Marconi, and Plessey. The list of American manufacturers was longer. Among the larger ones: Bendix, General Electric, Hazeltine, Hughes Aircraft, ITT Gilfillan, Raytheon, Sanders Associates, Sperry, Texas Instruments, and Westinghouse.

Radar-Guided Missiles Arrive

In the postwar era after the Berlin Airlift, radar applications multiplied while cold war tensions built up, boiling over into combat during the Korean War. U.S. military leaders sought better ways to

*German air traffic controllers
use a system called TracView.
PC technology plays an
integral part in the
functioning of the system,
bringing affordable com-
puting power to the user.*

*Test pilot Bart
Warren
flanked by the
family of
Hughes Falcon
air-to-air
missiles.*

defend against enemy bomber threats against the continental United
States, and in areas of combat. The answer lay in air-to-air fire-
control radar and radar-guided missiles.

The first air-to-air fire-control radar, which enabled a pilot to fire
at a target he could not see, was delivered to the U.S. Air Force in
1949. It was the Hughes E-1 (or APG-33 in military nomenclature).
It saw combat service during the Korean conflict and was in widespread use for many years thereafter. Mounted aboard USAF interceptor aircraft, the APG-33 and improved successors advanced the

technology and the military utility of airborne radars. The U.S. Air Force called it a pioneer achievement.

The Hughes Aircraft Falcon was the first air-to-air guided missile produced. With the Falcon and a radar fire-control system on board, the crew of an air force interceptor had a potent new combination. The radar fire-control system enabled the crew to locate enemy aircraft at ranges up to 30 nautical miles (NM) and "illuminate" it with radar energy whose reflection was detectable by a homing head on the missile. The crew could engage enemy aircraft at distances well beyond gun range, guiding the missile to hit and destroy the enemy. This contrasted with earlier systems, which relied on larger and more cumbersome rockets equipped with proximity fuses to explode their warheads near the target aircraft.

The Falcon missile went into service in 1954. It was a semiactive missile; the launch aircraft's fire-control radar continued to illuminate the target aircraft, and the Falcon's own seeker homed in on the pulses reflected off the target. The Falcon family of missiles (USAF designation AIM-4) served for more than two decades, seeing constant upgrade and improvement of its capabilities, including improved radar guidance and the addition of infrared detectors.

Air Force crews flying interceptor aircraft of the North American Air Defense Command were armed with Falcon missiles as they maintained the defense of the continent in the late 1950s and early 1960s. The Sparrow, or AIM-7, radar-guided missile built by Raytheon succeeded the Falcon and has seen three and one-half decades of service. It was introduced in 1958 and still serves in the 1990s. Like the Falcon, the Sparrow is semiactive; its internal radar homes on the target illuminated by the fire-control radar in its launch aircraft. Its range is five times greater than the Falcon's; it can engage targets at nearly 15 nautical miles (NM). Sparrow missiles accounted for 50 air-to-air MiG kills by U.S. Air Force crews in the Vietnam War, more than any other missile. During the Gulf War, Sparrow missiles registered 29 of the 41 confirmed Iraqi aircraft killed in air-to-air combat, again more than any other missile.

By the late 1960s, radar and computing technologies had advanced to produce the Phoenix air-to-air missile, the next logical step. Phoenix, designated AIM-54 in military parlance, was built by Hughes Aircraft with a much longer range than its predecessors (out to more than 100 NM) when it entered service with the Navy in 1974 aboard the F-14 Tomcat interceptor aircraft.

A U.S. Army soldier inside an operations shelter pinpoints the location of an enemy artillery position based on data provided by Firefinder radar.

Like the Sparrow and Falcon, the Phoenix was semiactive for part of its mission. During the first part, it was guided toward the distant target by the electromagnetic illumination from the main fire-control radar of its launch aircraft. But, unlike its predecessors, on the final dash over about 10 NM to its target, Phoenix used its own internal active radar transmitter for illumination. That capability for independent terminal guidance, matched with more computing power in the F-14's fire-control system, allowed the Tomcat crew to launch six Phoenix missiles against six targets simultaneously.

The Phoenix was a bridge between eras. It was a semiactive missile homing on energy from its host aircraft for most of the way, and then active for the terminal phase of its mission, using its own internal transmitter. Because it was semiactive, the pilot of the launch aircraft still had to keep his airplane and its radar in the vicinity of the target. That made them vulnerable to enemy action and countermeasures.

What the Air Force and Navy really wanted for the post-Vietnam era was a missile that did the work itself, giving them a true "launch and leave" capability. That meant the pilot could engage multiple targets or take evasive maneuvers after launching the missile, knowing the missile's internal radar system and computers were tracking and guiding the missile to a kill. The military also wanted the missile to be small, so planes could carry more of them.

The requirement was met by the Hughes Aircraft AMRAAM, or AIM-120 missile system. Entering service in the early 1990s, the AMRAAM has a self-contained radar and an inertial navigation system. It can be launched from very long ranges (up to 100 NM) and receive inertial data from its host airplane until it comes within radar range (said to be up to 60 NM). At that point, its own radar takes over and guides the missile to a kill. With AMRAAM, the pilot can launch one or many missiles at very long range, assigning each to a target, and then immediately turn away to evade or attack more targets. Once launched, AMRAAM missiles sort themselves out and attack individual preassigned targets. AMRAAM is the pinnacle of missile radar packaging in the early 1990s.

On the ground, radar systems were developed for detecting enemy movement. Several ground radar sets were used by U.S. Army and Marine fighting forces in Vietnam to provide alerts of enemy movement beyond eyeball range and in darkness and bad weather.

U.S. Navy infrared images (below) taken from sensors on a TC-4C training aircraft showed geologists significant thermal regions on Mount St. Helens, Washington, shortly after the May 18, 1980 eruption.

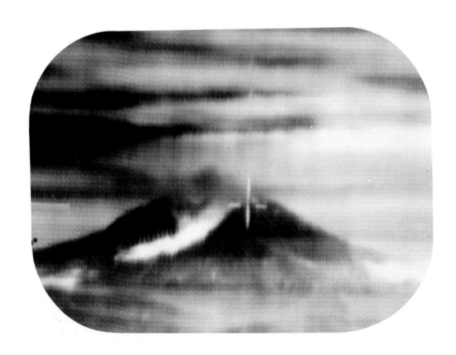

The radar's range was short, less than five kilometers, but that was much farther than troops could see, and thus a real combat benefit.

Other ground radar sets, such as the AN/TPQ-37 Firefinder, were developed to detect incoming enemy mortar or artillery projectiles and, by appropriate processing and analysis, to rapidly provide an accurate fix on the tubes that fired them. This counterbattery radar capability proved popular with ground forces protected by it during the Gulf War. It also tended to have a deterrent effect on enemy gunners, who found themselves swamped by steel fragments within one or two minutes of firing at U.S. forces.

The success of early radar led to immediate searches for countermeasures, ways to jam it or render it ineffective. Beginning in World War II and continuing through Korea, antiradar countermeasures were developed by nations as rapidly as radar improvements appeared. It was warfare by electrons; a serious contest.

The radar countermeasures contest appears in civilian as well as military life. Many police departments use small radar sets to detect drivers exceeding speed limits. Drivers in turn mount radar warning receivers ("Fuzzbusters") in their autos to be alerted that the police radar beam is scanning them.

More useful ways of using microwave radar in automobiles seem certain to appear on the market as the devices become smaller, cheaper, and more available. For example, microwave radar systems can enable drivers to cruise at selected speeds, yet remain safe distances from cars ahead, or they can provide warning in blind spots, helping to prevent mishaps.

Piercing the Darkness

Radar is one aid to seeing in the dark; the technologies grouped under the generic name of infrared (IR) are others. On the electromagnetic spectrum, the radar frequencies are in the high end of the radio range, roughly from 1 GHz to 100 GHz. The infrared frequencies are next above on the spectrum, situated between the radar frequencies and the frequencies of visible light.

Nature has provided certain reptiles with sensors that use the infrared frequencies. Snakes such as the boa constrictor, the python, and the sidewinder are able to seek out and pounce on other creatures at night because they have heat-sensing pits on the front of their faces. Sensors in these pits detect the heat of potential prey, guiding the predator snake unerringly to the victim. Like the earlier example

A U.S. Air Force weapons crew member checks an AIM-9L Sidewinder Missile mounted on an F-16 fighter during Operation Desert Storm.

of the bat, these are cases of nature equipping animals with an additional sensor useful for their survival.

Thermal-imaging applications are based on the knowledge that all objects in our environment emit heat to greater or lesser degrees and produce waves of infrared energy. Objects in space also emit heat in varying degrees, the Sun being a very hot example nearby. Even cold asteroids are detectable against the cold of the void, where the temperature is absolute zero. The infrared (IR) radiation from stars is especially useful, because it can be picked up with the appropriate detectors. Although stars are invisible to the human eye in daylight, their infrared radiation makes them stand out brightly against their background. The first star trackers were introduced in 1948, successfully tracking bright red stars in daylight.

Scientists and their military customers recognized quickly that infrared systems could be valuable adjuncts to airborne radar. They are passive, emitting no energy, so infrared search-and-track sets would not alert an enemy to their presence. They could find and track targets while avoiding detection by enemy radar. Also, since they operated at frequencies outside the radar range, infrared sensors were immune from electronic countermeasures.

The revolutionary A-12/YF-12 interceptors (later the SR-71 Blackbird) used the air-to-air infrared search-and-track system. In service in the early 1960s, it could detect a high-flying aircraft at distances up to 120 nautical miles.

But national authorities soon concluded that attacking bombers were more likely to approach targets at low rather than high altitude. That presented the air defense command with a real problem. The radar of the early 1960s was ineffective at detecting low-flying targets. To cope with the emergency, the USAF fleet of interceptor aircraft was fitted with infrared search-and-track equipment that could do the job.

Later in the 1960s, when the Surveyor spacecraft were sent on reconnaissance missions to the Moon, they were equipped both with radar and with an infrared star-tracking sensor for navigation. When the Apollo series of spacecraft journeyed to the Moon and returned, they carried an infrared docking device for the Lunar Lander and the Lunar Orbiter as backup for the radar.

The principle of a target's emitting stronger IR waves than its background led to the Super Falcon missile equipped with an infrared seeker, which appeared in 1959 for use on interceptor aircraft. The Super Falcon homed on the heat from a target aircraft's engine exhaust, literally flying right up its tailpipe and destroying it.

At roughly the same time, the Sidewinder (AIM-9) air-to-air infrared guided missile was developed; it has been the standard for more than 35 years. Like the Falcon, the Sidewinder has been upgraded over the years and will be used into the next century, clear evidence of its adaptability. The ninth version of the Sidewinder missile, designated AIM-9R, went into production in 1990. Sidewinder missiles were credited with 10 air-to-air kills in the Gulf War.

Development of infrared systems for military use was accelerated by the Vietnam War. Both Texas Instruments and Hughes Aircraft produced Forward-Looking Infrared (FLIR) systems that saw service in the conflict and were improved rapidly in the 1970s and 1980s. FLIR devices were installed on airplanes such as the OV-10 Bronco forward air controller aircraft, and for reconnaissance on

The B-1B prototype is equipped with state-of-the-art Forward-Looking Infrared (FLIR) sensors to produce a television-like image of terrain for the crew in nearly any weather condition.

a variety of helicopters and gunships. Later, advanced Hughes FLIR systems were fitted aboard B-52 and B-1B bombers, enhancing their capabilities for night and low-level operations.

Through the 1970s and 1980s, the use of FLIR systems on aircraft expanded in all the armed services. For ground combat applications, laser rangefinders and thermal viewers were mounted on tracked vehicles such as the M1 Abrams tank and M2/M3 Bradley Fighting Vehicles, enabling the crews to operate in darkness, bad weather, and through the smoke and haze of combat.

From the earliest days of space-based defense systems, infrared devices played a central role. They were able to detect the exhaust plumes of ground-based missiles as they launched and to detect targets in space. Infrared sensors and processors were also placed aboard satellites such as the Geo-Operational Environmental Satellite (GOES), providing data for weather pictures; and Earth observation sensors such as multispectral scanners and thematic mappers aboard the Landsat series of satellites.

Infrared Trends

The trends in infrared, or thermal imaging, in the 1990s are toward lower cost, better performance, and smaller size. As signal processing and display capabilities improved during the 1970s and 1980s, so did the characteristics of the substances used for infrared detection. The detectors in most military and space systems are made of mercury cadmium telluride (HgCdTe). In use, they are cooled to very low temperatures, near 77 degrees Kelvin (-196 degrees Centigrade). The trend was toward developing and exploiting other substances that operate effectively at higher temperatures. The eventual goal, close at hand in the early 1990s, is to create infrared detectors that require no cooling and are effective at ambient temperatures.

Another trend is toward packing more efficient detectors into ever-smaller space. As that occurs, the amount of information they generate climbs sharply. In 1992, an advanced infrared sensor setup called a focal plane array generates as much data as the words in the *Encyclopedia Britannica* — 13 times every second!

With the improvements in detecting and processing information, the costs of producing infrared systems has gone down. As prices drop and the technology becomes available commercially, the public sector will benefit from the research done for the military. Infrared sensor arrays could be mounted on autos, for example. Coupled with appro-

President George Bush checks out the advanced Tactical Weapons Sight, a compact yet powerful advance in thermal imaging for military and law enforcement applications.

priate displays, drivers could navigate through darkness, fog, or haze at daylight speeds. Robert L. Sendall, a pioneer in thermal imaging, suggests these additional applications among the many that seem feasible: aircraft landing aids, highway monitoring, commercial trucking, law enforcement, security surveillance, environmental monitoring, private automobiles, and the obvious one: fire detection.

Part II
Applying the Technologies

*Combinations of enabling technologies
create innovation, and then more innovation*

Galileo spacecraft atop its two-stage inertial upper stage after release from space shuttle, and before it sets off on its interplanetary voyage.

Chapter 7
See Near, See Far . . .
Then See Farther

**Electronic technologies enable humans to
see both farther and nearer, from images of
the outer reaches of space to exploring
inside the human body.**

I nfrared and radar technologies have enabled us to see what's
coming over the next hill, but the microscope and telescope
have provided a way for us to see ourselves and the heavens
around us with our own eyes. Advances in microscopes and telescopes
steadily have increased our ability to explore both near and far,
making it possible to focus on a single atom. While microscope and
telescope technologies of the past did not enable us to see below the
surface or beyond the line of sight, remote sensors, microprocessors,
and computer-generated imaging have extended the horizons of
scientific exploration to include everything from the inside of the
human body to the outer reaches of space.

Few would contest the role of the microscope in making the world
a healthier place. Until diseases could be identified and isolated in
the laboratory under microscopic examination, cures were a hit-or-
miss affair. Indeed, the world owes a significant debt to Dutch
scientist Anton van Leeuwenhoek, who announced his invention of
the world's first simple optical microscope in 1673. This began a
process that now enables scientists to see menaces to health that were
once too small to be discerned.

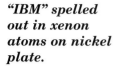

"IBM" spelled out in xenon atoms on nickel plate.

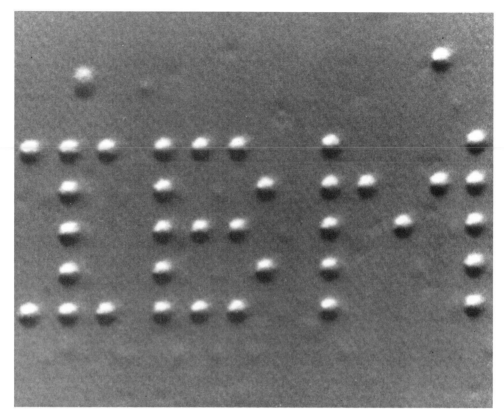

Advances in electron microscopes have moved rapidly. The first such microscope was invented by Ernst Ruska of Heidelberg, Germany, just before World War II; it had a magnification power of 12,000. Postwar improvements extended the reach of electronic microscope technology. Two IBM researchers, Gerd Binning of West Germany and Heinrich Rohrer of Switzerland, earned a Nobel Prize in 1986 for developing a scanning tunneling microscope (STM) that photographed atoms one at a time. An atom is only 3 angstroms wide (one angstrom is the equivalent of one ten-billionth of a meter). By 1990, scientists at the IBM Almaden Research Center in California, using an advanced STM, were able to pick up xenon atoms and move them around one by one to fashion an I-B-M design on a surface of nickel. Paul Hansma of the University of California-Santa Barbara sees STM technology as just beginning. "In fifty or one hundred years," he said, "the scanning tunneling microscope will be known primarily for the microscopes it inspired."

If seeing minute organisms otherwise invisible to the naked eye advanced medicine, seeing farther with greater clarity through telescopes brought about comparable and fundamental changes in global and space exploration. The telescope was invented by another Dutch scientist, Hans Lippershey, in 1608, one year before Galileo built his first model, which eventually reached a power of about 30, enabling him to confirm that the Sun is the center of the solar system. Until the telescope came into widespread use, discovery of new stars and greater understanding of the solar system was, at best, a function of 20-20 vision. It is no accident that the invention of the telescope and the opening of the Age of Astronomy occurred almost simultaneously. When Galileo searched the skies with a telescope in 1609, he made more discoveries in a single year than all the astronomers who preceded him had made in their lifetimes.

Antique optical instruments and modern optical microscopes and telescopes are all limited by line of sight. They can be focused on an object with varying degrees of magnification, but they cannot see what is hidden beneath the object's surface. The first time the surface beneath human skin was seen without breaking the skin occurred in 1895, when Wilhelm Roentgen, who was investigating materials that fluoresce when exposed to cathode rays, held his hand between a machine emitting X-rays and a photographic plate. The resulting gray image opened a new field for medicine. Within a few weeks, doctors started looking inside human bodies, and today X-rays remain a mainstay of medical diagnosis. X-rays cannot distinguish between soft tissues, however, and they expose the patient to radiation. More, standard X-rays provide only a two-dimensional image.

Experiments in the 1950s and 1960s proved that X-rays penetrated objects at different speeds, varying with the density of the objects, but the equations describing the differences were so complicated that only a computer could re-create the object in three dimensions. Two physicists, Allan MacLeon Cormack and Godfrey N. Hounsfield, invented a piece of equipment that could perform such a task. Computed tomography (CT) provided a picture 100 times as sensitive as the X-ray machines then used in hospitals.

The first CT scanner was installed in Wimbledon, England, in 1971, revealing secrets of the brain invisible to X-rays. Expensive CT scanners soon became available commercially in limited numbers. Later models were improved to map an entire human body by dividing the anatomy into 35,000 squares, each 1.5 millimeters square. The scanner requires only two seconds to take a picture of a given square. The system uses 1,000 detectors, each recording 2,000 measurements. Parallel computer processing made the system pos-

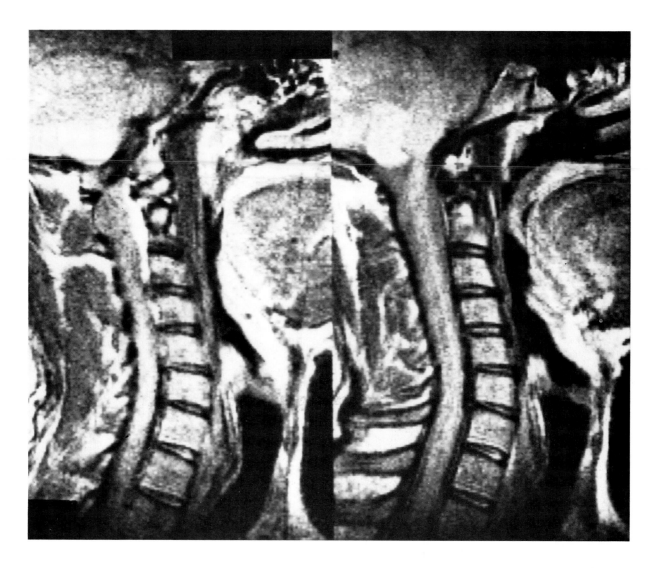

By using MRI images, physicians can assess nerve and disc damage to the cervical spine. Here the vertebrae are mostly normal and in good shape.

sible and effective, performing calculations concurrently that would otherwise be done sequentially and take much longer.

Sonogram technology, which incorporates a computer that compares the intensity of sound waves as they bounce back when transmitted through a portion of a body, is similar to CT technology. The echo principles of sonar are utilized in ultrasound equipment. Ultrasound is commonplace in obstetric clinics, but interpretation of the data is more art than science. One expectant mother was told by her doctor: "There are so many arms and legs thrashing around in there, I can't count how many babies you'll have." Twins were later delivered, much to the anxious mother's relief.

Ultrasound equipment embedded in a catheter (sometimes called acoustic microscopes) and inserted into arteries can reveal blockages.

Orthopedic surgeons, using technologies based on satellite-mapping of terrain, are able to map cartilage and bone in three dimensions, a significant advancement of the state of the art.

Magnetic resonance imaging (MRI) created another powerful tool to enable physicians to see inside the human body. Also known as nuclear magnetic resonance, MRI exploits radio waves and high-powered computing to create images impossible to devise by other methods. Raymond V. Damadian applied for the first MRI patent in 1972, and the devices were used first in Great Britain.

The operating principle of MRI is based on the knowledge that the nuclei of hydrogen atoms within a magnetic field change their alignment when bombarded with radio energy. When the radio waves are turned off, the atoms fall back into alignment and send off an electrical signal. Since the human body has many hydrogen atoms in blood and tissue, it is a natural arena for exploiting the MRI effect

Surveyor on the beach near Culver City, California. Surveyor explorers made five soft landings on the Moon, sensing and sending back invaluable data on the Moon and its surface.

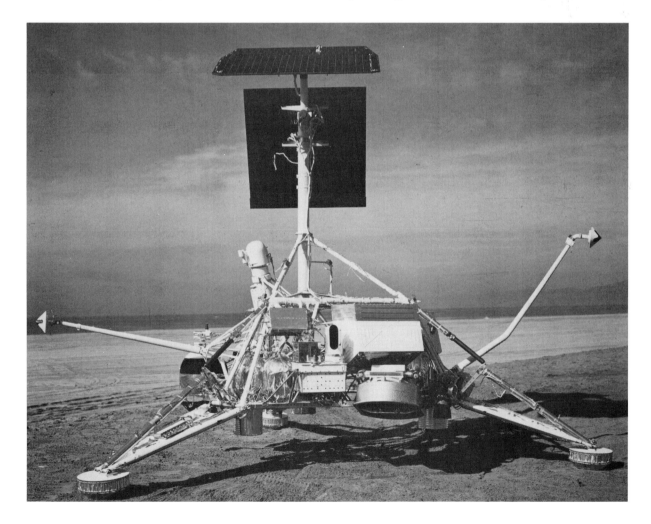

From space, satellites can monitor the receding tropical rain forests in South America. Here an area has been cleared for agricultural experimentation.

if the tiny electrical emissions can be detected, processed, and displayed. The MRI system surrounds the body within a structure that creates the radio frequency waves. That structure is part of a larger surrounding structure that creates the necessary magnetic field. The rig is linked to a computer and software that process the electrical impulses into high-resolution images for diagnosis.

By 1990, more than 2,000 such machines were operating in the United States alone. Magnetic resonance has advantages over CT scans. The MRI process allows a doctor to examine a part of the body from any angle. The procedure requires the patient to be motionless. One Washington, D.C., student recalls being the guinea pig during his physician father's first fitful attempt to operate a new machine: "He told me I couldn't move, because if you move the image gets ruined. I had to sit still the whole afternoon." Magnetic resonance has

Las Vegas, Nevada and vicinity, as seen from LANDSAT.

better contrast resolution between diseased and normal tissue than CT or ultrasound. And physicians can detect anomalies earlier. MRI has proved useful in discovering tumors that are too small to be detected by other means.

All these systems of noninvasive exploration are basically exercises in mapping the human body, a process sometimes called anatomic cartography. Similar linkages of electronic technologies have created advances in mapping the Earth and in reaching out to create maps of the planets.

Seeing Farther and Farther

Sensor technologies and computing power similar to those in MRIs and CTs are useful not only in mapping the solar system from Earth, but also in exploring the Earth from space, and in seeing what exists in the farthest reaches of space. Remote sensing provides an invisible electronic link between a satellite-borne sensor and the images it conveys to a ground-based observer.

Remote sensing was derived from military applications that began with aerial photography in World Wars I and II. It was

The GOES satellite flies over the equator and relays weather information to a ground station at Wallops Island, Virginia. This drawing depicts the area of useful cloud information relayed by GOES.

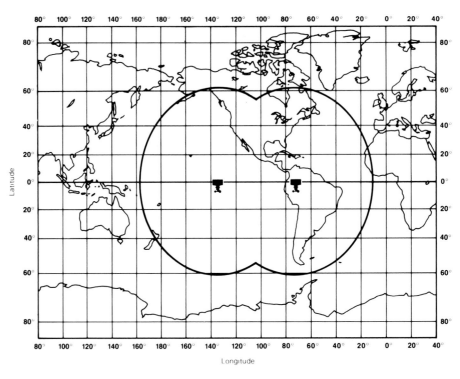

Area of Useful Cloud Information From GOES Satellites.

extended by satellite collection when the space age arrived. Although remote sensing usually refers to information gathered by satellites in space and returned to Earth electronically, it can also mean transmitting information acoustically through the water from unmanned submarines mapping the ocean floor or searching for wrecks.

    Sensors include one or a combination of devices that gather information across the spectrum of electromagnetic radiation. The

The space shuttle orbiter payload bay is open and the shuttle is about to release the rest of its load into orbit. A remote sensor camera aboard a co-orbiting satellite took the photo.

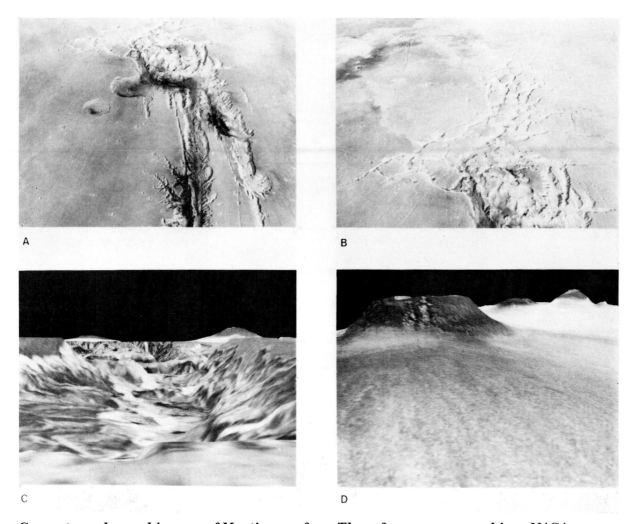

Computer-enhanced images of Martian surface. These frames were used in a NASA production entitled Mars the Movie. *The animated journey explores the Martian Grand Canyon as well as the highest points of the Martian surface.*

kind of sensor used depends on the purpose. Passive microwave scanners, for example, can see through clouds and can operate in either daylight or darkness because they sense emitted rather than reflected solar radiation. Accordingly, they are often used for weather data collection and forecasting.

Remote sensing was used in America's first environmental satellite, LANDSAT-1, which was launched in 1972 and immediately opened up new ways of looking at Earth. LANDSAT-1 was unique because it carried the world's first multispectral scanner, along with three television cameras. The scanner operated in different regions of the electromagnetic spectrum, and the television cameras in the

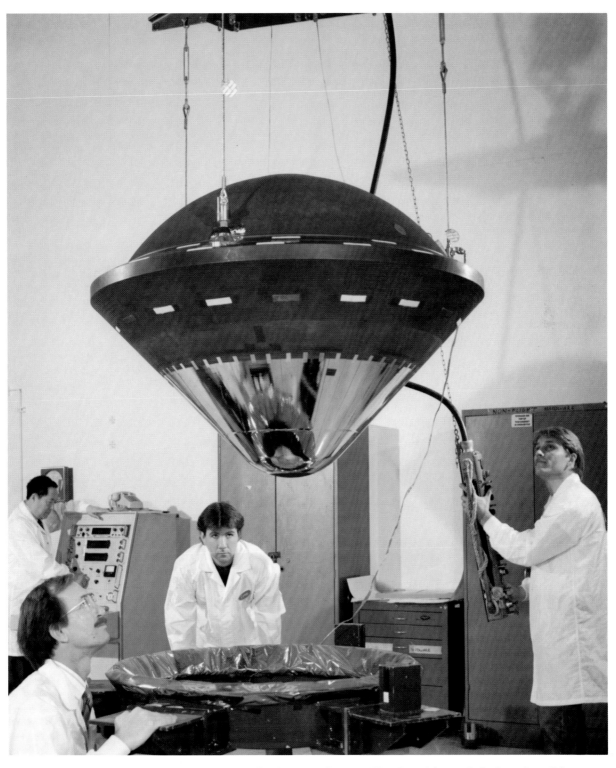

Technicians flank the Galileo probe during testing at Hughes Aircraft before its shipment to NASA for integration with its spacecraft, built by JPL. The Galileo probe is expected to enter the atmosphere of Jupiter in late 1995.

visible-light range. The scanner represented a new technology, so the initial proposal by Hughes Aircraft to include it on the satellite produced a controversy that ended only after the satellite was launched. Within a month, the three TV cameras all malfunctioned and went blind. However, the multispectral scanner continued operating dependably. Although designed to last one year, it was still functioning a decade later, mapping areas the size of New Jersey in less than 30 seconds each time it takes an image.

Remote sensing can be used in oceanography, meteorology, climatology, natural disaster prediction, water resource and pollution studies, cartography, engineering, and geology. Satellite images may even lead anthropologists and archaeologists to locations for excavation that were undiscovered until detected by advanced sensors aboard satellites.

Remote sensors were used in TIROS 1 (Television and Infrared Observation Satellite), the first of four separate satellite programs (totaling 10 satellites) designed to gather meteorological information. Launched on April 1, 1960, it was covered with 9,000 silicon solar cells producing 28 volts, sufficient to power wide- and narrow-angle television cameras and a continuous infrared scanner. Over a 78-day experimental period, TIROS 1 returned more than 22,000 images, about 60 percent of them usable.

Images provided by satellites have proven of enormous value. Weather satellites are the most visible evidence of success. Their products are widely broadcast by television stations around the globe. For some of the world's population, early weather predictions can mean life or death when a disaster is in the offing. Early warning allows relief agencies time to prepare appropriate remedies. Predictions of a 1992 drought throughout eastern and southern Africa caused planners in the U.S. Agency for International Development (AID) to revise their estimates for food grants. Some experts familiar with the capabilities of weather satellites point out that the weather trends leading to famines of the past could have been predicted and many lives saved.

Satellite imagery enables geologists to examine Earth's structural formations in their totality, predicting movement in surface and subsurface formations. Far-infrared imaging can locate hidden underground streams and rivers, information that is useful to engineers interested in structural foundations, helpful for tracing the underground movement of pollutants, and practical for finding water

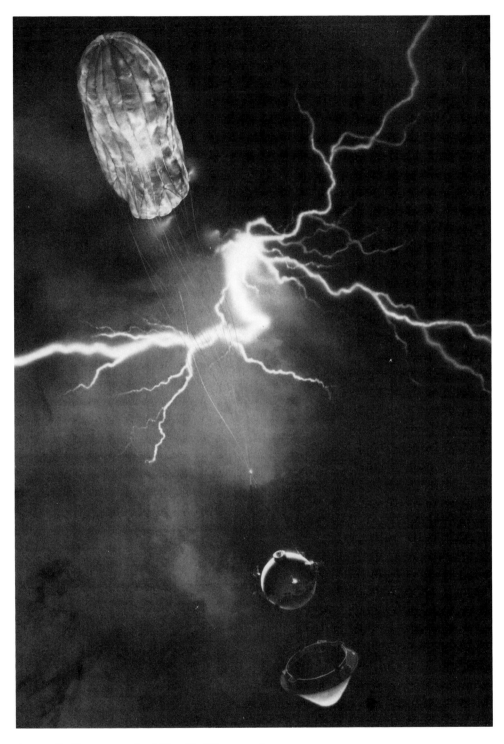

*An artist's concept of the Galileo probe
entering the atmosphere of Jupiter.*

sources during drought. Other imaging procedures can locate geo-thermal sources or help monitor volcano activity. In May 1980, a Navy aircraft flying past Mount St. Helens in Washington State and equipped with thermal-imaging sensors first detected the heat emissions from the awakening volcano, enabling authorities to verify its imminent eruption and to evacuate citizens from the immediate area in time.

Other satellite information gathered from a variety of sensors allows better response time to natural disasters, such as earthquakes, hurricanes, and forest fires. Oil spills and pipeline ruptures can also be detected and analyzed. Because satellites gather so much information, the problem is often too much data, not a shortage. It will require advances in computer memory capacity and innovative processing to handle the deluge.

Processing the Information

While integrating advanced technologies makes it possible to look beneath the surface of the human skin and to see the Earth in new ways from satellites, the one technology that crosses all lines is the microchip that handles the complicated tasks of managing space travel. The microchip played a crucial role in the Pioneer Venus project. Pioneer Venus was launched from the United States on May 20, 1978, reached Venus on December 4, 1978, and immediately began sending back photographs from an orbit around the planet ranging from 241 kilometers (180 miles) to 66,000 kilometers (41,000 miles). On December 10, three different probes launched from the Pioneer Venus orbiter began their descent to the surface of the planet. Scientists analyzing the information soon learned that Venus contains about 300,000 times as much carbon dioxide as the Earth (the ultimate greenhouse effect?) and no apparent water whatsoever, certainly less than 0.1 percent in the atmosphere.

In 1990, radar sensors on the Magellan probe orbiting Venus detected and sent back to Earth an incredible amount of detail about Earth's sister planet. Robert McD. Adams, secretary of the Smithsonian Institution in Washington, D.C., said the flood of data from Venus (and other planets) has created an entirely new field, comparative planetary geomorphology.

An earlier space voyager, Pioneer 10, carried the first microprocessor to be flown in space, an Intel 4004, to manage and control a mass spectrometer. According to NASA officials, the microprocessor made it possible to take a reading once every kilometer (0.625 mile) change in altitude. Without it, a reading of only once every 10 kilometers (6.25 miles) would have been possible.

Pioneer 10 celebrated its 20th anniversary of space travel on March 2, 1992, at which time it had reached a distance eight billion kilometers (five billion miles) from Earth. Pioneer 10 carries messages on a small plaque in case it eventually falls into the hands of intelligent beings. Included on the plaque is a drawing of a man and woman. Originally the man and woman were clasping hands, but it was decided that an extraterrestrial, not familiar with a human figure, might conclude that the two were really one being connected at the hands, possessing four legs, two heads, and two outer arms, so the idea was dropped. In about 32,000 years, Pioneer 10 will reach its first star, Ross 248. The spacecraft has the distinction of being the human artifact that has traveled the farthest from the Earth.

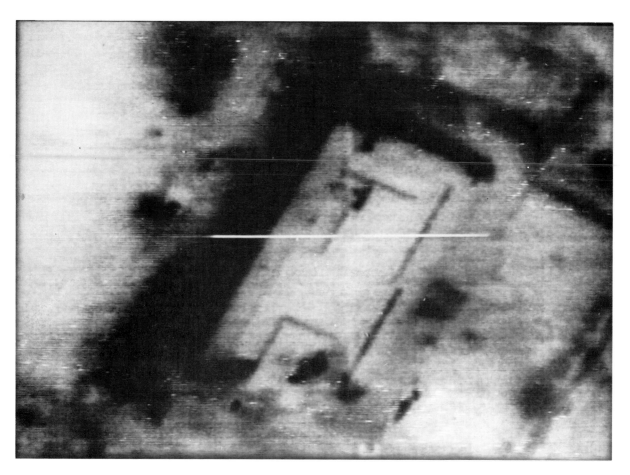

A video camera in the nose of a "smart" bomb delivered by an F-117 "Stealth" fighter of the 37th Tactical Fighter Wing shows the bomb homing in on a Baghdad communications center, its crosshairs on the center of a camouflaged command post bunker during Operation Desert Storm, January 18, 1991.

Chapter 8

Putting It All Together

As the longbow revolutionized warfare in the Middle Ages, so did electronic technologies create smarter weapons that changed warfighting in the late 20th century.

T hrough the centuries technology has transformed warfare, as the Gulf War of 1991 demonstrated so dramatically. In that war, electronics were integrated into radically new systems that changed the nature of war just as significantly as the longbow, the machine gun, and the airplane had done in earlier eras.

When the longbow was introduced to medieval warfare, it increased the distance between combatants. Instead of battling each other face to face, warriors engaged in deadly combat from opposite sides of the field. An English archer of the day could shoot 12 arrows per minute, to a lethal range of 227 meters (240 yards). At Crécy in 1346 and Poitiers in 1356, and again at Agincourt in 1415, English archers defeated French armies by using longbows made of yew wood that stood as tall as they. The barrier of arrows put up by the English archers was lethal against both mounted and foot troops. The longbow technology may not have made war more civilized, but it changed the ways that generals maneuvered their armies.

Centuries later, the long rifles of the American revolutionaries were used to deadly effect in battles from Lexington to Yorktown. They forced the British to change their battle formations from massed squares of packed soldiers lest they suffer severe losses. In later wars, the machine gun and the airplane forced comparable

Image of Lambert Field, St. Louis, from prototype of synthetic array radar in F-15 Strike Eagle at a distance of 80 miles. Its resolution is analogous to a person's being able to see a postage stamp at a distance the length of a football field in the darkest night, in a driving rain.

changes in military operations. Then, as now, new technology provided more precise and powerful tools of war.

The Gulf War of 1991 witnessed extensive reliance on electronic technologies at every level of operations. The technologies were integrated into systems that may have astonished the onlooking world, but did not surprise the scientists and engineers who developed them, nor the young men and women who used them.

The high-tech systems wielded against Iraq had their beginnings nearly two decades before, when the Pentagon's Office of Defense Research and Engineering provided the funds and direction for what proved to be crucial basic and applied research. Money was allocated for large-scale integrated circuits for high-powered computing, for advances in thermal imaging to enable sustained combat at night, and for research and engineering on technologies to make aircraft nearly invisible to radar.

At the most basic level, microprocessors were ubiquitous in the Gulf campaign. They were embedded in all military weaponry. The military commands relied on advanced microprocessor applications in devices such as laptop computers and fax machines that were carried in the field. Microprocessor applications were invaluable for sorting out the massive logistical buildup. They were especially useful in integrating every one of the flights anywhere in the combat area into a single daily master plan that contained complicated routes, altitudes, airspeeds, and missions.

The sea and land wars were integrated with the air campaign through computer systems developed especially for that purpose. For controlling the sea war, navy admirals used an Advanced Combat Director System (ACDS) that absorbed information from sensors, ships and aircraft, and intelligence sources. This was integrated into automatic tracking that displayed information according to users' requirements, leading finally to evaluation of enemy threats, air control intercept orders, and automatic firing of appropriate weapons against enemy targets. Just a week after Desert Storm began, Bill Rediker of ABC News reported that "aboard the Aegis cruiser *Valley Forge,* the most sophisticated surveillance system in the navy has been tracking the allied campaign against the Republican Guard."

Satellites Everywhere

Communications satellites served both military and commercial needs during the crisis that began with Iraq's invasion of Kuwait on August 2, 1990. Amateur video images taped during the invasion were flashed via satellite from Kuwait City to television audiences worldwide. As the crisis built, national leaders as well as private citizens received essential information from images conveyed over satellites by broadcasters such as the Cable News Network (CNN) and the British Broadcasting System (BBC).

Two views of the same Iraqi airfield in close-up. The top was taken in December 1990; the bottom was taken two months and several air strikes later.

Saddam Hussein used this extraordinary access to global audiences to play his own game of cat and mouse. For a time, he held scores of hostages from several nations and ensured that their detention and release were reported by the television networks with the aim of influencing opinion in the affected countries.

On the night the shooting war began (January 16-17, 1991), three CNN newsmen and their camera crews transmitted images via satellite of Iraqi antiaircraft tracers arcing through the black sky, trying to shoot down attacking F-117 Nighthawk aircraft that were invisible to them. Days later, after Iraq had captured several downed airmen, Saddam displayed them on television for the global audience. After first ejecting news organizations from the coalition countries, Saddam permitted CNN to station reporter Peter Arnett in Baghdad with a portable television transmitter that could send video images via a satellite uplink. As the air campaign progressed, Arnett's video transmissions, including interviews with Saddam, were a unique source of information for both military and civilian audiences. Saddam himself relied on satellite news broadcasts such as CNN's to receive information about the world outside Iraq.

The coalition forces relied heavily on the Military Satellite Communications Systems (MILSATCOM), commercial systems such as International Telecommunications Satellite Organization and INMARSAT, and satellite systems belonging to allied nations. The Defense Satellite Communications System carried much of the military message traffic.

On patrol at dusk near the Iraqi border, 101st Airborne soldiers check their position with a Global Positioning System receiver.

Planning and operational information came from other satellites. Weather reports from the first days were drawn from information provided by the Defense Meteorological Satellite Program (DMSP). Civilian systems also served as sources of data, but public distribution of the information was curtailed at government request

for the duration of the conflict to keep vital weather data out of Iraqi hands. Throughout the campaign, imagery from Earth observation satellites such as LANDSAT and the French-owned SPOT augmented that from the military satellites.

For navigation, satellites provided position information of unprecedented accuracy. The NAVSTAR global positioning system (GPS) enabled individuals and units to determine their location to within a few meters. The 16 NAVSTAR satellites in low-Earth orbit sent data to portable receivers in the hands of troops on the ground, in the air, and at sea. The GPS receivers were equipped with powerful microprocessors that took signals from the satellites and converted them into an immediate display of the receiver's latitude, longitude, and altitude. For the first time in combat, soldiers and their leaders did not have to ask, "Where am I?" There was no excuse for getting lost, even in the trackless desert.

The U.S. Navy used GPS to help guide Tomahawk cruise missiles to their targets with precision. GPS was also used to help keep ships in the Persian Gulf away from known minefields. The precision of the GPS system has thousands of commercial applications as well. It was used in January 1986 by ships in the Atlantic salvaging debris from the space shuttle *Challenger* disaster to pinpoint the areas to search. By 1992, commercial GPS receivers were in widespread use in general aviation and recreational boating, as well as in agriculture. Farmers mounted GPS receivers aboard their heavy equipment to provide them with precise position information for applying fertilizers.

Lasers Abounded

All the services in the campaign took advantage of laser technologies built into their weapons. The Hellfire missile, launched from Army AH-64 Apache helicopters, struck targets with pinpoint precision because it used a laser seeker for guidance. A laser designator on the Apache directed a beam of energy onto the precise spot to be hit. The laser seeker in the Hellfire's nose homed in on the spot and destroyed the target. U.S. Air Force, Navy, and Marine Corps units employed their laser weapons in the same fashion.

Laser rangefinders used by U.S. tank crews also demonstrated the effectiveness of integrated electronic high technology. Tank crewmen used the laser to instantly determine the distance and direction to a moving Iraqi target, as well as the target's speed, enabling the gunner to destroy the target while his own tank kept moving. According to General H. Norman Schwarzkopf, "Our [M1A1 tank] sights have worked fantastically well in their ability to acquire targets though all the dust and haze."

Lasers and infrared detectors were utilized both to achieve and to prevent surprise. At the tank battle of 73 Easting, visibility on the battlefield ranged from 200 to 1,300 meters. U.S. Army cavalry troopers had an advantage over the Iraqi adversaries, however, because the M1A1 tank gunners used an advanced thermal imaging sight that could detect Iraqi tanks through the dust to ranges of 3,000 meters. To see an Iraqi tank was to kill it.

Surviving Iraqi troops said that fire from the U.S. tanks came as a complete surprise. They thought they were under attack from aircraft flying unseen in the clouds above, so they leaped from the vehicles and took up positions in the sand, where they faced the choice of surrender or death. Most surrendered.

Radar's Many Roles

Radar in its multiple applications was almost as omnipresent as microprocessors. A main strategic goal was disabling Iraq's radar and communications networks to "blind" Saddam Hussein and his generals so they could not communicate with each other or with their fighting forces. The air campaign was devised accordingly. With its electronic eyes and ears shut down, Iraq's air defense guns, missiles, and aircraft were hard-pressed to defeat the allied air assault.

The air battle itself was managed from modified Boeing 707 four-engine jet transports with the military designation of E-3A AWACS (Airborne Warning and Control System). AWACS aircraft were aloft around the clock, flying in prearranged flight paths in designated sectors. The downward-looking radar antenna mounted in a saucer-like rotodome atop the fuselage detected aircraft out to a range of 400 miles. The information was processed by high-speed computers and sent to mission controllers seated in the cabin at consoles with video displays. The controllers monitored the air situation, sorting out friend from foe and providing instructions for friendly aircraft.

The role played by the AWACS in the first air-to-air kill of the war illustrates how important airborne radar is in modern combat. During the night of January 16-17, 1991, Air Force Captain Steve Tate was aloft leading a flight of four F-15C Eagle fighters. He and the others were hunting Iraqi fighters that might interfere with coalition

Mission controllers at battle stations in the AWACS E-3A during Operation Desert Storm. Battlefield coordination was made remarkably easier by the AWACS and JSTARS systems.

air operations. They were but one small segment of the 668 aircraft in the aerial armada over Iraq that night. Their onboard pulse-Doppler radar sets enabled them to detect and track small, high-speed targets at altitudes down to treetop level.

While Tate and his comrades roared through the night toward Baghdad, three AWACS aircraft flew racetrack patterns high above. Their rotodomes turned at six revolutions per minute, scanning the battle area and tracking every aircraft in flight. Suddenly, AWACS controllers radioed to Tate that an unidentified aircraft was in his vicinity, flying about 12 miles behind. Tate and his wingman swung around and maneuvered into position behind the target. He used his radar to "lock on" to the target and identified it as an Iraqi Mirage

Opposite, M1 tank gunner sights a target through its thermal imaging system, which turns night into day.

F-1 flying about 16 miles in front of him. He called out "Fox One" to indicate he was firing an AIM-7 radar-guided Sparrow missile. The Sparrow dropped off his wing, its rocket motor ignited, and its internal radar was guided to the target by energy from the radar in Tate's aircraft. Within a few seconds the missile slammed into the Iraqi Mirage, setting it on fire. Captain Tate had scored the first aerial strike of the war.

On succeeding days and nights of the 43-day air campaign, AWACS radar was used to control 3,000 aircraft flights per day, a density higher than the busiest airport in the world, Chicago's O'Hare International. In fact, control was so precise that not a single midair mishap occurred during the thousands of combat flights.

Three Air Force crewmen carry a 155-pound AIM-9L Sidewinder heat-seeking missile to its mounting position on board an Air Force F-16.

Other advanced communications and surveillance systems, such as the aircraft that housed the Joint Surveillance and Target Attack Radar System (JSTARS), achieved their most widespread use after the offensive phase of the war began. The JSTARS radar detected slow-moving ground targets such as trucks and tanks and Scud missile launchers and fixed their position precisely.

Information from AWACS and JSTARS was added to information from other intelligence sources to provide snapshots of the movements of Iraqi forces. Long-range surveillance radars in high-flying TR-1 and U-2 airplanes provided instant images of activities hundreds of miles distant and transmitted them to the command headquarters for evaluation. Even higher in the sky, KH-11 satellites flying at altitudes of 120 to 300 miles collected images that were immediately available on the ground. Men of the special operations forces, meanwhile, infiltrated deep into Iraq to collect intelligence information providing confirmation of information collected by the high-flying sensors in satellites and aircraft.

Advanced multimode radar systems in F-15E Strike Eagles provided pictures so detailed that they rivaled black-and-white photographs. One pilot said that the radar was so good that from 40 to 50 miles away (60 to 80 kilometers) "you could take a picture of your housing tract, then not only pick out your house, but tell if your car was in the driveway."

Radar was an essential component of the Patriot missile's performance defending against Iraqi Scud ballistic missiles during Desert Storm. Patriot crews were helped by warnings from satellite-based infrared sensors that detected exhaust from Iraqi Scud missiles. Detection information was sent simultaneously to an Australian ground station and Cheyenne Mountain at U.S. Air Force Space Command in Colorado Springs. Computers at both locations instantly verified the launch, predicted an impact area, and alerted Patriot crews in the Gulf region. The key to the Patriot system, which is produced by Raytheon, is a radar system that on its own searched for, detected, and tracked an aerial target, while quizzing the unidentified target to determine if it was friend or foe. Trained troops at an engagement control center then made the decision whether to launch the missiles.

The U.S. Army's AN/TPQ-37 Firefinder radar starred in the counter-fire mission. The Firefinder system, built by Hughes Aircraft, used a large stationary antenna whose phased-array radar sent a rapid sequence of beams along the horizon, forming an electronic radar curtain spanning 90 degrees. Any enemy projectile penetrating the electronic curtain was immediately detected, and the information sent to advanced signal-processing software. The location of the

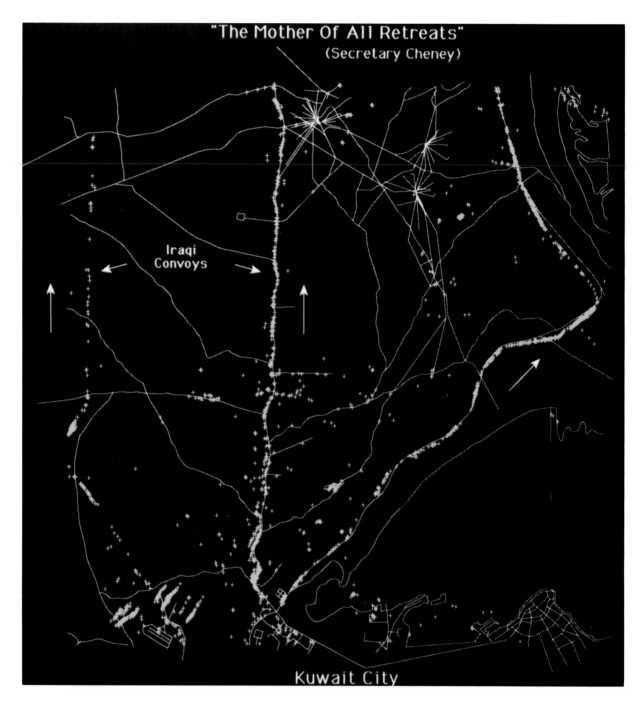

The "Mother of all Retreats," as it has been called by military officials. Iraqi convoys flee from Kuwait in this JSTARS real-time photo.

enemy firing position was quickly determined to within 10 to 15 meters before the projectile had hit the ground.

The enemy location was displayed immediately on a large map and on video screens and at the same time sent automatically to fire-direction computers of artillery units. The information automatically adjusted friendly cannon to fire accurately at the spot from which the enemy projectile came. Iraqi gunners soon learned the risk of unleashing their guns.

Infrared: Night and Day

Infrared sensors literally made the difference between night and day for coalition forces in the Gulf War. They stripped the cloak of darkness from Iraqi forces, making them as vulnerable as if they were operating in full daylight.

The offensive phase of Desert Storm began with a nighttime attack by eight U.S. Army AH-64 Apache helicopters. Their mission: to knock out two critical Iraqi radar sites that were links in a chain of radar stations that provided early warning to Baghdad of aerial intrusions. The Apache pilots used Forward-Looking Infrared (FLIR) sensors for navigation and to detect the radar sites, which appeared on the display screens in the Apache cockpits. "This one's for you, Saddam," shouted one of the pilots, Chief Warrant Officer 3 Dave Jones, as he sent his laser-guided precision Hellfire missiles speeding toward their targets. FLIR sensors in the helicopters recorded the event on video tape. Briefing the news media later, General H. Norman Schwarzkopf said the Hellfire missiles "plucked out the eyes of Iraq's air defense early warning system" in less than five minutes, creating a radar gap that opened the doors to 100 strike aircraft.

On that first night and for the remainder of the campaign, F-117 Nighthawk Stealth aircraft executed strikes on high-priority targets in Baghdad itself. The Nighthawks exploited their onboard infrared sensors, their near-invisibility to radar, and radio silence to slip through the strong Iraqi air defense radar network and the surface-to-air missiles and antiaircraft artillery defending the Iraqi capital. On board the F-117s, two infrared night vision systems looked forward and downward, producing daytime-quality pictures for the pilots.

The Maverick missile was another precision weapon used by strike aircraft such as the F-15E and A-10 Warthog. Maverick could be fitted with any one of three types of seekers for guidance: electro-

An avionics technician checks the APG-65 radar of an F/A-18 Hornet aircraft.

optical television, infrared, or laser. This is how they were used in combat. The two-man crew of an F-15E Strike Eagle armed with Mavericks, for example, located an assigned target while still some miles away. They designated the intended target by placing a cursor over its image on one of the multimode screen displays in the cockpit. They then relayed the information via computer linkup to the Maverick missile carried aboard the aircraft. The Maverick's seeker locked onto the target before launch. When the target came within missile range, the missile was launched, and the crew members kept their cursor on the desired point of impact. The seeker in the

Maverick's nose flew to the aim point. The accuracy was astounding. As one U.S. Air Force pilot said: "Wherever you aim the Maverick, it hits the target." Another version of the Maverick in development has a millimeter-wave, high-frequency radar seeker that will enable it to hit targets autonomously. The pilot of the launch aircraft can fire it and forget about it; the missile will do the work.

The A-10 Warthog attack aircraft did not have its own FLIR system, yet was required to operate at night, especially to seek out and destroy Iraqi Scud missile launchers. Innovative A-10 pilots figured out how to turn on the Maverick missile's infrared seeker and use it for nighttime navigation. The image created by the IR seeker was displayed in the cockpit; while intended for guidance to a target, the pilots used it for navigation and terrain avoidance by scanning a narrow swath of the terrain ahead of them. It was not perfect, being compared with peeking through a keyhole into a room, but it gave the pilots a much-needed nighttime capability.

Maverick missiles were used with deadly accuracy during Operation Desert Storm. This missile has been one of the most successful military weapons ever deployed.

The U.S. Navy also used infrared technology in its air-to-ground missiles to attack high-priority targets. Navy A-7E Corsair II attack planes carried Standoff Land Attack Missiles (SLAM) that combined the AGM-84 Harpoon antiship missile containing a proven global positioning guidance system, with an onboard infrared seeker. This integration of technologies made SLAM almost a fire-and-forget weapon with surgical strike capability against land or sea targets.

After the SLAM missile was launched, its positioning system collected information from GPS satellites and passed it to the missile's inertial navigation system, bringing the missile within sight of its target. The missile's imaging infrared seeker, which was turned on as the missile approached the target, sent a video image to the aircraft controlling the missile, where a crew member identified the target and locked the missile onto its deadly course. In the meantime, the Corsair that launched the SLAM had turned and evaded the target area, and was up to 50 miles away.

Optical Guided Weapons

Optical guidance was applied to several weapons employed to good effect during the campaign. For example, video sensors from the Maverick missile were mounted on GBU-15 guided bombs. After launch from an aircraft, the sensors in the bombs transmitted pictures of the target back to a cockpit television monitor where a crew member controlled the missile's flight to strike a precise location.

Many television pictures of targets being hit during Desert Storm came from equipment in these weapons, which were later used to stop the flow of oil into the Gulf from an oil pipeline off the Kuwait coast. Iraqi forces had opened the pipeline to damage the Kuwaiti oil industry and, in the process, began environmental warfare. Oil industry experts were consulted on how the damage might be contained and the flow stopped. They recommended bombing a small valve assembly called a manifold that was situated in the pipeline control area, believing that if it were disabled the oil flow would stop. Achieving this required high precision. Crews of F-111 Aardvark aircraft from the 48th Tactical Fighter Wing launched GBU-15 2,000-pound bombs equipped with Maverick electro-optical seekers against the small targets. En route to the target, the GBU-15s transmitted video images of their flight to the F-111 crews. The GBU-15s struck precisely, and the oil flow was stopped.

After the successful strike against the pipeline valve assembly, General Schwarzkopf narrated the videotapes himself in a briefing for the press and the television audience. "What you're going to be doing," he told the audience, "is looking directly through the TV camera [Maverick seeker] in the nose of that munition as . . . it's flown right into the small manifold area to destroy it."

The TOW (Tube-launched, Optically tracked, Wire-guided) was another optically tracked missile that saw widespread use in the campaign. The TOW entered army inventories in 1970 and saw first combat late in the Vietnam War against North Vietnamese Army tanks. In one battle, a single U.S. Marine destroyed 10 tanks with the TOW system, and an army aviator killed 17 enemy tanks with TOW missiles launched from his UH-1 Huey helicopter.

In the Gulf War, much-improved TOW missiles were precise tank killers launched from AH1-S Cobra helicopters, Bradley Fighting Vehicles, and the High-Mobility Multipurpose Wheeled Vehicle (nicknamed Hummer). Army and marine gunners firing TOW missiles destroyed tanks or other targets at ranges up to 3,750 meters.

TOW-armed helicopters worked especially well. "Anything we saw, we engaged with TOW missiles," one U.S. Marine Corps pilot recalled. "They pack quite a wallop. And anything that we hit, we destroyed."

Value of Precision Weapons

Precision weapons such as infrared and laser-guided missiles accounted for seven percent of the total bomb tonnage dropped during Desert Storm. But that small percentage accounted for about 80 percent of the targets destroyed.

They helped end the combat sooner and with less loss of life and equipment. Perhaps the greatest savings, however, came from being able to destroy an intended target with one weapon on the first attempt. By contrast, during Vietnam, according to U.S. Air Force estimates, it took more than five bombs to be assured of hitting the target within a few hundred feet, and during World War II it required more than 500 bombs to achieve accuracy within thousands of feet.

Implications for the Future

Technologies have come a long way since the longbow, and there is little reason to believe the trend toward greater sophistication will stop or even slow down, given the pace of new developments.

These changes mean that the need for brains in the military may now be as important as that for brawn. Muscle has utility, to be sure, and modern warriors must be physically able to take the pace of combat. Yet warriors of the future will not be selected primarily for their physiques; even today, soldiers are selected primarily for their intelligence, because smarter soldiers are more easily trained, are more goal-oriented, and are likely to have fewer discipline problems. Army recruiting programs now focus on brighter youths who are at least high school graduates, are ambitious, and are eager to succeed.

If the longbow increased the distance between warriors, new electronic technologies made destruction possible without any physical contact between opposing sides. Combatants have become increasingly divorced from their targets to the point that the enemy is no more than a signal on a scope. As reliance on technology increases, even infantrymen are less likely to encounter the enemy face to face.

The GBU-12 laser guided bomb was successfully deployed during Operation Desert Storm. Smart weapons like this were able to hit their targets with pinpoint accuracy.

TOW 2A

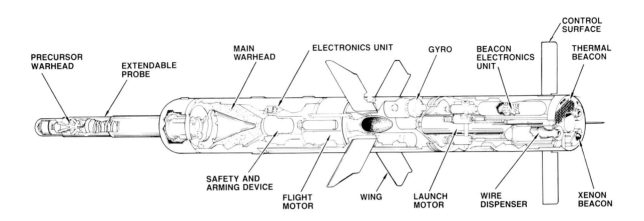

Desert Storm saw U.S. tanks equipped with electronic sighting devices destroy Iraqi tanks invisible to the naked eye, a far cry from the days when soldiers faced one another's cold steel. Fewer soldiers with better technology will be able to do more with less. Fewer people will be affected when nations mobilize, perhaps making it easier for policy makers to resort to force, even as the ability to command and control from the highest to the lowest levels becomes easier thanks to the communications revolution. These developments may influence the process by which a nation selects its military leaders. Where special technical needs are perceived, some military leaders may be chosen more for their technical competence than their tactical savvy.

The major components of the TOW 2A antitank missile.

Inevitably, armies, navies, and air forces will see their logistics and maintenance support structures (the "tail," in military slang) grow as the size of their fighting forces (the "teeth") diminishes. But the greater demands to keep systems operational will call for more specialization in maintenance, supply, and support.

Will this affect the cult of the warrior? There will always be a demand for people of great daring willing to undertake difficult and dangerous missions, but the days of massive forces may fade. Still, given uneven development, particularly in the Third World, low-tech warfare will be around for some time. The ultimate challenge for the United States and other highly developed nations is to use science and technology to field forces that are capable of both low- and high-tech warfare for as long as potential threats remain in the world.

Students in isolated outback areas of Australia can interact with the nation's best teachers in urban centers from remote locations through satellite programming carried via satellites such as Aussat.

Chapter 9

Revolutionizing Communications

New technologies fostered the Information Age and now meet its communications needs.

I n the Information Age of the 1990s, new computer and communications technologies have changed how information is created, processed, and distributed. This revolution in communications has freed people from the constraints of time and space, enabling information to reach them freely and permitting them to be in touch with others whenever they choose. Consider the simple transaction of calling another person in a distant city. To make a telephone call to another city in 1950, you dialed up the long-distance operator and gave her the name of the city and the telephone number you wanted to reach. The long-distance operator, in turn, contacted one or more additional operators along the path until the linkages between your number and the distant one were made.

In its advertising of that year, the Bell System proudly proclaimed that its operators could get your call through in as little as a minute and a half. In 1950, the nation had 250,000 hard-working telephone operators to put calls through, and they processed 6,200,000 long-distance calls and handled part of the 164,400,000 local calls made each day.

When telephone planners of the 1950s looked ahead at population growth and potential demand for telephone services, they report-

The number of operators declined dramatically as the number of phone calls increased exponentially. Modern switching technology and computer capabilities made this possible.

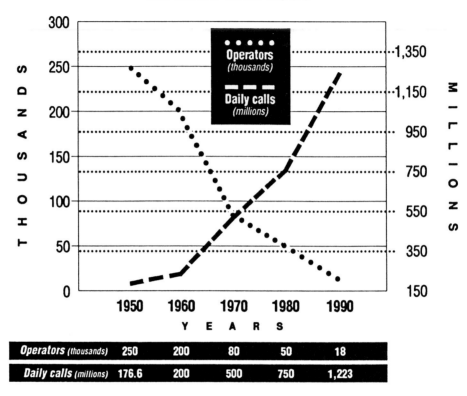

Telephone Operators and Phone Calls

In 1950 and 1990

Operators (thousands)	250	200	80	50	18
Daily calls (millions)	176.6	200	500	750	1,223

edly projected that within 20 years, based on existing equipment and anticipated volume of traffic, all of the working women in the United States would have to be telephone operators.

Communications technologies altered these projections dramatically. By 1990, the number of long-distance telephone operators nationwide had shrunk to 18,000, about 7 percent of the 1950 figure. At the same time, the number of daily local and long-distance calls exploded to 1.226 billion, more than 7.2 times those made every day in 1950. By 1990, a call from a touch-tone phone in the United States to anywhere in the country and almost anywhere in the world could be made without the intervention of an operator and was completed in less than half a minute.

The limitations of time have been overcome in other ways too, thanks to technology. In 1992, if the distant party doesn't answer the telephone when you call, you can still convey the information you wish by leaving a message on the party's answering machine or voice mail. The information is stored and accessible to the person called whenever he or she taps into the system to retrieve it. In return, that person can call you back when convenient and leave information for you. Impossible in 1950, but commonplace in the 1990s.

The restrictions of space were also overcome by new technologies. In 1950, if you wished to make a telephone call, you had to be at a fixed point—a location that had a telephone connected by wires and cables to the national system. Governments, the military, police forces, and big businesses had radio networks and radiotelephones, but their numbers were small and their capacity limited. By 1992, however, a person wishing to make a telephone call could *be* just about anywhere and place a call *to* just about anywhere else via cellular telephone. Even if you were not a cellular subscriber, you had access to cellular phones. Auto rental companies at airports were among the first to offer such phones for rent in their cars. On airlines, telephones have become available in seat backs or at convenient spots near the restrooms or galleys. Sir Edmund Hillary, the first man to scale Mount Everest (May 1953), told a Washington audience in May 1992 about receiving a telephone call at his home in New Zealand not long before. His son was calling from the summit of Mount Everest, using a portable telephone!

As recently as the early 1980s, mobile phones of any size were considered rich men's toys. However, by early 1992, nearly eight million subscribers were on cellular phone service in North America alone. These subscribers accounted for slightly more than half the world total of 15.4 million. The cellular technology was embraced rapidly, once prices were low enough and accessibility adequate. Hughes Network Systems noted that cellular phones in North America reached the 7.5 million level in less than four years. Comparable market penetration for the personal computer took almost seven years, and comparable market share for television receivers took 13 years to achieve.

Ubiquitous, Interactive, and Personal

Examples abound of the transformations produced by the revolution in telecommunications: instant images of events appearing on television screens worldwide; election results tabulated immediately and transmitted within minutes after the polls close; financial

The Early Bird communications satellite undergoes a checkout before its launch in 1965.

exchanges open and operating around the clock. Availability of cash from a checking account used to be restricted by both space and time — the bank's location and its schedule. Nowadays, with the appropriate card, you can get cash or make transactions around the clock at automatic teller machines.

The availability of communications along with computer memory and processing power contributed to the credit card explosion in the 1970s and 1980s. In 1980 there were 86.1 million cardholders nationwide using 526 million cards. By 1989, the numbers had grown to 109.5 million and 956.9 million, respectively. The projection by the Nilson Report for the year 2000 foresees continued, but more modest growth, to 118.3 million cardholders and 1.167 billion cards. The cards are usable anywhere, any time, because merchants can verify their authenticity via data links. The funds can be transferred between accounts immediately over similar links.

The blending of several technologies helped accelerate transformation in communications. Advances in transmission were integrated with storage, processing, and display technologies to create a totally new world of communications where a person need never be out of touch with the flow of information. Communications satellites and fiber optic cables built new capabilities atop the existing global communications system and were married with computer chips and laser light to create entirely new businesses. Direct marketing, for example, has expanded rapidly thanks to the blending of electronic technologies. Companies such as L. L. Bean in Freeport, Maine, and Lands' End in Dodgeville, Wisconsin, were founded in communities far from major urban centers. But they were able to prosper and grow into multimillion-dollar enterprises by skillful use of communications and computers.

Video teleconferencing is another useful business communications tool made possible by combinations of technology. Using satellite or fiber optic cable connections and large-scale displays, people in one location can see and communicate with another individual or group in one or more distant locations.

The coupling of technology and computers has freed businesses from the constraints of space. Businesses using brain power instead

The Growth of Electronic Mail
Public & Private Mailboxes (U.S. Only)

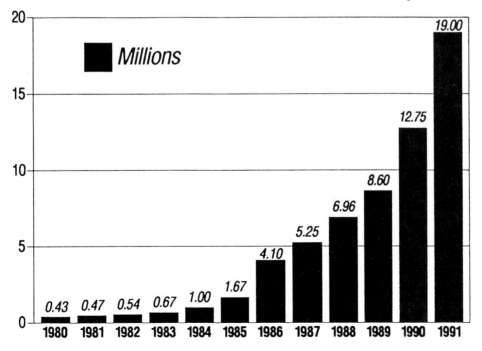

The number of electronic mailboxes grew exponentially in little more than a decade.

of production processes have become decentralized; people can work from their homes or from small offices away from headquarters while linked electronically to associates, customers, and suppliers. Indeed, cellular phone companies note that about half their customers are self-employed and one-quarter work for companies with fewer than 100 employees.

With the new technologies, larger companies with far-flung operations can decentralize while ensuring access to essential information. Businesses with heavy internal communications requirements can install a private satellite network for less than the cost of leased telephone lines. One of the users of these instant private networks, the Wal-Mart retailing chain, reports that with a satellite the company can communicate with all its stores at once. This makes possible the sharing of merchandising information as well as corporate-wide training.

Cellular communications technologies will soon bring about the "wireless city."

Using electronic data interchange, businesses can also refine the processes of buying supplies and selling products, reducing costs and space required for inventory.

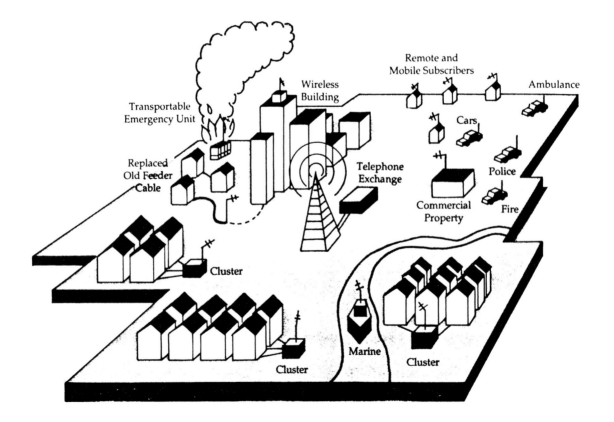

Freedom from time and space is also afforded by facsimile machines and electronic mail. Even into the late 1970s, fax machines were slow and expensive; it took six minutes to transmit a page, and each machine cost several thousand dollars. With the advent of microprocessors and multiple communications channels, along with sharply lower prices, fax machines became ubiquitous at the beginning of the 1990s. At the same time, the use of electronic mail (E-mail) burgeoned, with mailbox holders surging from zero in the mid-1980s to 19 million by 1991.

The capacity of communications systems is being expanded by digital technology. The cellular phone system was based on analog technologies since its introduction in 1983. But beginning in 1992, digital cellular systems became the standard of the future. By using advanced technologies, cellular channels can handle up to fifteen private calls per channel simultaneously, compared with a single call on an analog channel. The audio quality of digital cellular communications is comparable to analog.

The advent of digital cellular technologies has reached the point where the term "wireless city" has meaning. Putting the term into practice, GTE Telephone Operations and the USTC subsidiary of International Mobile Machines (IMM) will select a town or cluster of towns in rural Texas and upgrade the telephone system from wire to digital radio (diagram opposite). Stuart Brand, author of *The Media Lab: Inventing the Future at MIT,* said, "With digitalization, all of the media become translatable into each other — computer bits migrate merrily — and they escape from their traditional means of transmission . . . if that's not revolution enough, with digitalization the content becomes totally plastic — any message, sound, or image may be edited from anything into anything else." Digitalization created the potential for multimedia applications that began to enter the marketplace in the early 1990s.

Business Applications

Today, computers exchange data with other computers over high-speed data links. Video teleconferences conducted over satellite nets save travel time and money. Educational programs and professional consultations are routinely conducted over satellite channels. Most international finance is transacted via computerized communications by satellite. Maritime activities are tracked using computers and satellite relays. And late-breaking news flashes are sent around

Home subscribers with a compact 18-inch satellite antenna mounted on a windowsill will receive more than 100 television channels of news, sports, and entertainment programming.

the globe in an instant by satellite, bringing disparate peoples closer together.

The next thrust for entertainment-oriented satellites will be direct broadcast television to homes, which will open the door to the next revolution in home entertainment — high definition television (HDTV). Pictures as sharp as 35-millimeter film will be accompanied by symphony hall-quality sound. To achieve such high resolution, HDTV requires more of the radio frequency spectrum than TV stations or cable. Using advanced techniques for signal processing and for compressing signals, the desired higher resolutions can be achieved, beaming HDTV directly from satellite to home, where the signal is picked up on a small receiving antenna whose diameter is eighteen inches or smaller. Instead of creating acres of rooftops sprouting dish antennas, these smaller, more efficient antennas will be mounted unobtrusively on windowsills or in windows.

Video-transmitting satellites have advanced the cable TV industry and helped inaugurate a new programming form, narrowcasting. The rapid success of the cable networks has been due, in large part, to the diversity of special-interest programs. Viewers can go shopping, play a game, watch a movie, or take classes without ever leaving their living rooms.

In urban areas of the United States and in other advanced countries, fiber optic cable has become a popular medium for high-speed communications systems because of its capacity, security, and potential for future growth and expansion. The cost of fiber optic cable continues to drop, and installation techniques improve. Indeed, as fiber optic communications expanded in the early 1980s, the demand for copper dropped off temporarily.

The real potential of fiber communications will begin to be realized when fiber cables reach homes in large numbers. The TV Answer Co. of Reston, Virginia, began operations in 1992 based on a potential market for interactive transactions over cable television networks.

Down the road lies the wired city. The wires are already in place, in the form of either copper or fiber optic cables. From 1950 to 1989, the national network of telephone cables increased tenfold, from 155 million miles to 1,502 million miles. Fiber optic cables are the choice for future exploitation of the multimedia digitalization cited by Stuart Brand, because the light waves can carry so much more information than electrons through copper.

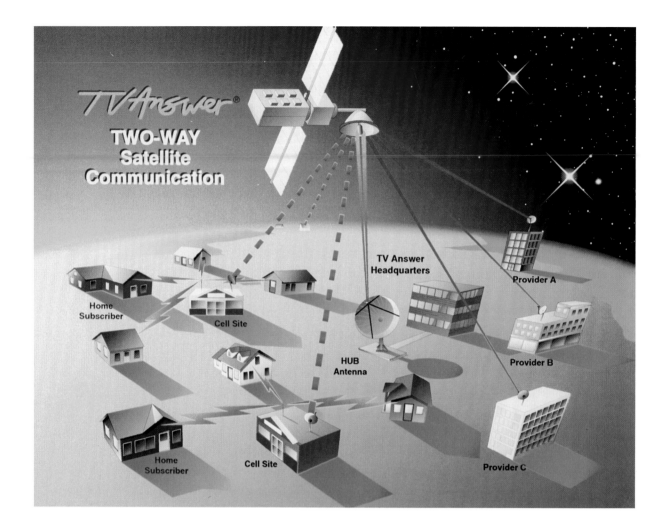

Communications network operations have wrought changes in medical practice as well. Over long distances, physicians can transmit to other physicians' images from X-rays, CAT scans, computer tomography, and magnetic resonance imaging. This lets physicians have on-the-spot interactive consultations with specialists and receive the best diagnosis from the best available resources, independent of distance.

Communications and Education

The new communications technologies have the potential to transform education systems. Universities, especially science and engineering departments, were quick to link up in networks to exchange information and to stay abreast of the field.

Educators saw early on the prospects for harnessing new communications technologies for "distant learning." Indeed, the state of South Carolina created one of the first statewide educational television networks back in 1965. Now, the South Carolina Educational Television system produces courses, conducts teleconferences and training programs statewide, and produces award-winning programs that air nationwide on public television. All told, 26 states operate educational TV systems.

In the state of Kentucky, the courts stimulated a bold experiment in statewide application of technology in education. In 1989, the Kentucky State Supreme Court declared the funding mechanism of the statewide school system unconstitutional because students from poorer districts were shortchanged. The justices directed the state legislature to redesign the funding mechanism. The result was the Kentucky Education Reform Act of 1990, a daring attempt to create statewide standards. The effort included integrating technology into education by building communications links among Kentucky's 176 school districts and the state department of education. When the system becomes fully operational, all classrooms will be equipped with computers, telephones, and video monitors that enable teachers to call up programs and information at any time.

The Kentucky system will make it possible for every student in the Bluegrass State to have equal access to the best education the state can provide.

The National Education Association believes that the rapidity of change in technologies requires adaptation in the nation's schools. But the NEA concurs in the judgment of the congressional Office of Technology Assessment (OTA), which says that "most teachers want to use technology, but few have found ways to exploit its full potential." The NEA argues that "technology cannot be an end in itself. The challenge then is to integrate the roles of teachers and technologies."

Opening Global Communications

With the rapid breakthroughs in technology, it is not surprising that engineers have arrived at ways to facilitate communications without wires or cables. Wireless technologies are beginning to replace outdated communications systems throughout the United States. Wireless communications represent a leapfrogging over old technologies and the old infrastructure of the wire telephone system.

Executives in USA teleconference with general in Brussels over PictureTel System 4000, Model 800.

Overseas there is enormous potential for completely new wireless systems to be built where communications systems are sparse or nonexistent or have been destroyed. In areas as diverse as the Commonwealth of Independent States, Eastern Europe, Africa, and the Pacific Rim, Australia, and New Zealand, the linkage of cellular, microwave, and satellite communications with personal computers is creating entirely new communications systems without the expense of laying cable or stringing miles of wires festooned on thousands of wooden poles.

Companies such as International Mobile Machines, BellSouth Enterprises, and Hughes Network Systems are among the leaders in opening these areas to total communications systems. Countries setting up communications with new technologies do not have to take the evolutionary steps followed by the United States, Canada, Great Britain, and many other developed nations. For example, Hughes Network Systems is creating the first digital communications system going into Russia, based on advanced digital technologies. It is in Tatarstan, a Vermont-sized region of Russia about 500 miles east of Moscow. The local system is on the level of rural America in the 1920s. Waiting times are extraordinary: several hours to call elsewhere within Russia, several days to call abroad, and a wait of up to 32 years for installation of a new telephone. The advanced system, to be operational in 1993, will be capable of tying into the creaky existing system, but will offer instant communications links within the region as well as to the world outside.

Revolutionizing communications frees people from the handcuffs of time and space. Rural and undeveloped areas become linked to the rest of the world. When you can communicate from anywhere, you can operate from anywhere and interact with everywhere else. This transformation is just beginning.

Looking ahead, the notion of a person's having a single personal communication number that is useable anywhere is being tested by companies such as BellSouth. The number would be unique and independent of area codes and exchanges. Indeed, if a single word could be selected to describe the trend in the communications revolution, it is access: access *from* an individual to information in many forms anywhere in the world, and access *to* an individual by others, from any place on the globe.

Hard-working F/A-18 aircrew members hone their skills in simulators that present real-time operational situations and emergencies.

Chapter 10

Inventing the Future

A blending of electronic technologies permits people to re-create battles or travel through space or time (while keeping one foot firmly on Earth), and it promises to revolutionize the future of industrial production.

I t is possible to invent and experience the future in virtual reality. Just ask Jack Thorpe. As the special assistant for simulation at the top level of the Defense Department, Thorpe makes planets, creates virtual worlds, and builds flying carpets and time machines.

Thorpe's magic carpets can convey people instantly about a battlefield or take them from an airplane cockpit to a tank turret with the click of a mouse key. Thorpe's time machines can set you inside an Iraqi tank during a battle of the Gulf War or place you in an airplane cockpit flying in a dogfight of the future. All of this has become possible through advances in simulation and technology.

Thorpe and his colleagues stimulate and provide the seed money for the research that makes simulation breakthroughs possible. Since the mid-1970s, Thorpe has seen simulation progress in extraordinary leaps.

It was once mainly a tool for training airplane pilots. Now, simulation encompasses much more. It links simulator systems around the globe that truly create the virtual worlds that Thorpe talks about. Simulation technologies allow humankind to invent the future. It was not always so.

World War II air crews trained on the Link Trainer. It was their introduction to instrument flying.

Beginnings of Simulation

When Orville and Wilbur Wright began to fly, no one was qualified to give them instruction. The Wright brothers learned by trial and error. In the process, they survived countless crackups. Orville was the pilot and sole survivor of the first fatal powered-airplane crash on September 17, 1908, at Fort Myer, Virginia. His passenger, Lieutenant Thomas Selfridge, was killed, becoming the first of many who would lose their lives in airplane crashes.

When others began to fly, they learned from the Wrights, from other self-taught aviators such as Glenn Curtiss, or on their own. In 1910, the Army had only one airplane (a Wright Flyer) and one pilot (Lieutenant Benjamin D. Foulois). Foulois taught himself to fly through correspondence with the Wrights, experimenting and learning with each flight, and seeking their advice after crashes and problems. This was not very good methodology. Foulois and other aviators passed on their techniques by instructing neophytes.

Early Military Simulators

When World War I came along, such casual practices were unsatisfactory for producing large numbers of competent aviators. The U.S. Air Service and its counterparts in the other warring countries began to formalize flight training. Among the devices they used were airplanes that had the wings clipped off. These "wingless wonders" taught fledgling aviators how to taxi aircraft on the ground. They were crude but effective simulators. Two-seated training aircraft, such as the ubiquitous Curtiss JN-4D Jenny, made instruction in the air possible. Aviators such as Jimmy Doolittle, Eddie Rickenbacker, and thousands more took instruction in Jennies during the war. In the postwar years, additional thousands of aviators such as Charles Lindbergh learned to fly in them.

But a true simulator did not come into being until the demands of World War II made it imperative. During the intense training of hundreds of thousands of young men, the U.S. Army Air Forces lost thousands of lives in mishaps. Without the crude simulators, the cost in lives would surely have been higher. One important simulator was the

Fires burn from the wreckage of an Iraqi tank destroyed by U.S. 2d Armored Cavalry in the Gulf War's battle of 73 Easting. This was the first battle to be re-created in simulation.

Link Trainer, a miniature aircraft replica with basic instruments and controls for the student pilot to observe and manipulate. The Link pivoted on its base so it could move in response to pilot manipulation of the controls. It could be covered with a hood so the student had no outside references. The student was forced to concentrate on the instruments and to learn to make his mind process and have confidence in the information the instruments were providing.

For bombardier training, other simulators were devised. They often were replicas of the bombardier's work station from an actual aircraft, such as the B-17 Flying Fortress or B-24 Liberator, complete with the Norden bombsight mounted on rails. The contraption rolled slowly on tracks above a painted countryside. Other crew members, including aerial gunners, practiced their deadly trade in simulators that were actual powered turrets propelled along railroad tracks, from which the gunners blazed away with their machine guns just as if they were customers at carnival shooting galleries trying to bring down models of enemy aircraft that flashed by. Maintenance crews practiced on mockups of the hydraulic or electrical systems of the aircraft they were to service.

But well into the 1970s the simulators were not part of a total training system. They did not portray the operating environment realistically. In fact, being inside the Link Trainer was like being in a dark cave. Nor were users capable of interacting with each other.

Simulation Payoffs

Even with their limitations and complications, simulators played a significant role in aviation training. And as more computational power became available, training simulators added capabilities. They permitted air crews to engage in maneuvers that would be impermissible in the air, including emergency situations from which a pilot might not be able to recover in real life. The big simulators were mainly the province of government and major airlines, and operating them was far less costly and hazardous than using high-performance aircraft. (This was also true for driver training simulators of the late 1960s and 1970s, which enabled students to encounter and deal with situations that could have been injurious or fatal if tried on the road.)

Training in simulators saved money and released real aircraft for missions, but there still was not enough computational power, display technology, or communications capability to realize simulators' operational potential. Slowly, the technology improved. In the 1960s, NASA provided simulators for the Lunar Lander and the X-15 hypersonic research aircraft. Both vehicles were venturing where humanity had not been before, and neither could be tried out incrementally. But by using simulators, Scott Crossfield and Bob White could practice operating the X-15 at the high speed of Mach 5.0. For the manned mission to the Moon, Neil Armstrong and Buzz Aldrin had practiced landing on the Moon and returning dozens of times before they ever lifted off from Cape Canaveral, Florida. The actual vehicle that was to take them to the surface was the lunar excursion module. In its simulator, built by Bell Systems, they saw accurate representations of the Moon's surface that familiarized them with what they would encounter on approach to landing.

Handling the Information

As users demanded more fidelity in simulators, powerful computer processing and massive computer memories were needed. Someone had to convert a picture of real life into computer language, figure out how to store and process it when needed, and display it so well that the user's brain interpreted the scene as nearly real. For a mission to the Moon, the players had only one chance to get it right, so money and software engineers were marshaled to get the job done.

But such largesse was not, and will not be, available for most simulation applications. As a consequence, the massive amount of information required to describe reality in computer code has to be reduced to a manageable and affordable amount.

At the low-cost end of the market, a scene is approximated by a number of polygons that appear on the screen, each appropriately oriented to simulate a three-dimensional feature. The user's brain makes up the difference.

In the intermediate cost range, actual dimensional photographs of the scene to be simulated are used. The illusion of three-dimensional scenes is created by using a computer to calculate what the photo would look like if it were printed on a rubber membrane that was stretched to conform to the actual 3-D scene. The computing load is reduced by having the membrane conform only at selected points. This approximation is like projecting portions of photographs on faces of polygons; lots of both, but still short of the best.

In the high end of the cost market, the closest representation of reality is achieved. For each elemental area of an image (and there may be tens of thousands in a small scene), the contribution of each one that lies along the line of sight joining the observer to the area is

Fidelity of simulation is demonstrated by comparing a photo of a TOW anti-tank missile fired from a Bradley fighting vehicle (above) and a simulated version of the same scenario (opposite).

calculated. Each one of the thousands changes as the observer's eyes change their scan, or as his aircraft or tank moves through space.

Linking Simulators

As simulators evolved, they benefited from improvements in computation and telecommunications. These advances made possible the creation of simulator networks that linked multiple simulators together. For example, the U.S. Army-Air Force SIMNET project, sponsored by the Defense Advanced Research Projects Agency (DARPA), began as a tank crew operational trainer. Then a wingman tank crew was added. The crews of both tanks could see the same scene, each on its own displays. The system was interactive, so actions by one crew were played out on both its own and its wingman's system. The system progressed until more and more tanks and even air support units were tied in together.

This Van-Dorn desktop industrial simulator replicates the characteristics of an injection molding machine on the production line.

Soon, Jack Thorpe's concept of using one system to train people at several locations came into being. Thus, pilots flying A-10 Warthog attack aircraft in simulators at an air base in Louisiana could interact with the five crews of an army tank platoon operating in their own M1 Abrams tank simulators at Fort Hood, Texas.

The payoff came during the buildup of Desert Shield and the combat of Desert Storm. In their years of training before the war, air crews for all U.S. services and coalition forces had used simulators to rehearse combat missions. U.S. Navy and Marine Corps aviators in F/A-18 Hornet attack aircraft flew countless combat missions in simulators before actually launching into the skies over Kuwait and Iraq, as did their counterparts in other aircraft and other services. Ground combat units, particularly the armored forces, reaped the benefits of realistic training they had received in simulators such as the SIMNET system.

In confrontations such as the Battle of 73 Easting, ground combat units harvested the rewards of training they had received in simulators. The battle has been re-created using simulation technology that reconstructs a gallant fight by three troops of armored cavalry — Eagle, Ghost, and Iron troops — on the afternoon and evening of February 26, 1991. The cavalry troopers fought and defeated a larger entrenched Iraqi armored force and cleared the way for the U.S. Army's VII Corps to destroy three Republican Guard armored divisions. The cavalrymen were mounted in M1A1 Abrams battle tanks and M3 Bradley Fighting Vehicles. Visibility by eye was limited to 200 to 1,400 meters because a sandstorm was blowing. But the U.S. vehicles were outfitted with thermal sights and laser rangefinders that enabled crews to see and to fight at distances out to 3,000 meters. Their fighting vision was two to 15 times that of the Iraqi troops in their Soviet-built tanks and carriers.

The cavalry troopers destroyed 50 Soviet-built T-72 and T-62 battle tanks, more than 35 other armored fighting vehicles, and 45 trucks. More than 600 Iraqi soldiers of the 12th Armored Division and Tawakalna Republican Guard Armored Division were killed or wounded, and at least that many were captured. The battle was recognized as a critical moment in the campaign.

Seizing the Opportunity

The information needed to reconstruct the battle was available in action logs, oral and written interviews, recordings from combat radio networks, recordings made by the participants during the battle, and overhead photography made before and after the battle. On the battle site itself, trained observers marked friendly and enemy positions. The mountain of information was processed and translated by computer software into a system that ultimately displayed the action precisely as it occurred. The simulation included a replication of the terrain and the weather (including the sandstorm), individual vehicle positions, and their movement about the battlefield. The re-creation enables viewers to see, feel, and understand the battle environment as never before. It allows for creating the simulation of "What if?" scenarios.

Simulation technologies and training programs built upon them are today showing promise for industry. In mid-1991, software and hardware developed for training crews of the B-2 Stealth bomber were modified for use in training people who operated the injection molding machines at the Inland Fisher Guide Co. in Anderson, Indiana. The training included maintenance and trouble-shooting of the equipment. The outcome: better-trained operators at considerable savings in machine and instructor time. At the same time, operator scores on functional tests improved.

The potential market for industrial uses of advanced simulation technologies is enormous but still largely untapped. As executives in manufacturing and service industries come to realize the capabilities of simulation, and the favorable cost-benefit ratios of using it in integrated training schemes, the market is bound to emerge and flourish. There is no question that simulation technologies can provide a new way to improve American industry.

It is possible to invent and experience the past, present, and future in virtual reality. Today, the term "virtual reality" is coming into common public usage after years of being the province of experts. Virtual reality refers both to the experience of being in an artificial environment and the medium that makes the experience possible. In a virtual reality, the user is included as part of the simulation.

Virtual reality differs from other computer graphics in two important ways, involving multiple senses and being interactive with the user. At the same time, virtual reality is similar to other applications of electronic technologies by integrating sensors, computing processing power, communication of information, and realistic displays. Indeed, virtual reality is the latest logical extension of the field of simulation.

It is the precursor of an industry about to burgeon as technologies of simulation become available to the entertainment business. It is now possible to put individuals into a two-seat "you-are-in-control" simulator for the leisure industry. Individuals can experience the sensations that have until now been enjoyed only by fighter pilots. In theme parks, for example, large simulators with up to 60 seats provide people with the collective experience of plunging over Niagara Falls or flying through the Grand Canyon.

An example of one rather basic sort of virtual reality is telepresence. It is the representation of the real world in real time, but it enables a person to experience being in two places at once. Project JASON, an activity to stimulate scientific interest in teachers and students, permits students at more than a score of remote locations to take part in expeditions to such places as the Galapagos Islands and the bottom of Lake Erie. Via interactive hookups, students and teachers can observe the action at the expedition site and, by adding telepresence, students far from the expedition can take control and operate unmanned submersibles exploring under water thousands of miles away. They are at home, but their telepresence operates the distant equipment.

Students at the U. of Rhode Island Graduate School of Oceanography Primary Interactive Network Site experience the JASON Project Voyage III to the Galapagos Islands.

NASA's Ames Research Center has drawn on telepresence. Astronauts equipped with a specially designed helmet can remain inside a space station, seeing things outside through a robot's sensors. As the astronaut's head turns, the robot's eyes swing simultaneously, enabling astronauts to assess situations and make repairs without leaving the station.

In a more complex use of simulation in a real-life teaching hospital, surgeons in training operate on simulated patients, learning from their mistakes instead of burying them. Or an operator at a nuclear reactor is projected into the heart of the reactor to assess damage and make repairs. The possibilities are limitless, and they are increasing with every new advance in simulation technology.

Part III
Looking to the Future

How has our world been transformed,
and where will we go from here?

Satellite technology reveals wind-flow patterns over the United States. The information network manager evaluates, controls, and keeps data flowing.

Chapter 11
Transformations

Widespread applications of electronics technologies have wrought profound transformations in everyday life and commerce.

T he world of the early 1990s is only four decades removed from 1950, the starting point for this book. Yet in terms of technological progress, the transformations that occurred from 1950 to 1993 is like a leap of four millennia. It is unprecedented in human experience.

When you bought clothes or groceries in 1950, the clerk wrote your purchases on a paper pad and rang them up on a manual cash register. Yes, a bell actually rang when the transaction was completed. You paid cash, or the amount was charged to your account at the store — if you had one. Stores closed down at least twice a year to take inventory; clerks counted each item by hand.

In 1992, the items you wish to purchase have bar-code labels. When you are ready to check out, the cashier passes a laser scanner over the codes. The purchases are totaled, and the inventory is automatically updated by computer. You may pay by cash, check, or credit card. If you are out of cash, chances are an automatic teller machine is in the store or close by so you can withdraw funds. If you pay by credit card, the checkout person "swipes" the card through an electronic scanner. The card is verified, the purchase recorded, and the transaction entered into your account, one of nearly two million such transactions made daily in the United States.

In 1950, banks were open from 9:00 or 9:30 a.m. until 2:00 p.m., weekdays only. One reason: the time required to post the transactions laboriously by hand.

In 1992, banks have extended their operations and their weekday hours and are routinely open on Saturday mornings. Your paycheck can be deposited electronically, and payments can be made electronically to places you have designated. You can use a bank credit card for purchases nearly anywhere in the world. The bank need not be open for you to withdraw cash; any convenient automatic teller machine will dispense it to you and debit your account immediately.

In 1950, parents in the United States wanting to talk from home with their sons in the Korean War zone had to make reservations days ahead, wait many minutes for the calls to go through, and then limit their conversations to a few minutes, even as they tried to hear the familiar but faint voice through the buzz of static.

By the time of the Gulf War in 1990-91, telephone calls between troops in the combat theater and their families at home were completed in a minute or less, and the conversations (routed via satellite and fiber optic cable) were as clear as if the other person were next door. Broadcasting organizations also arranged video conversations between home and the war zone.

In the early 1950s only governments, big science, and big business operated computers. The ENIAC of the late 1940s cost more than $3 million, weighed 30 tons, and processed 100,000 instructions per second. The IBM 701 of 1952 vintage was much smaller and cheaper and the fastest in the world, capable of processing 17,000 instructions per second. Before 1970, the average IBM data processing system had about 0.1 megabyte of main memory and contained up to 100 megabytes in on-line disk and tape storage.

In 1992, computers are ubiquitous in the home, office, and even pocket. Desktop computers using the Intel 486 microprocessor can process nearly 20 million instructions per second, about 1,200 times faster than the IBM 701, and costs $2,000 or less. As for computer memory, internal memories in desktops costing less than $2,000 can hold 120 megabytes of data. A single 1992 memory chip can itself hold as much as 16 megabytes of data, with 256-megabyte chips foreseen by 1997. On-line and disk storage memory capacity is essentially unlimited.

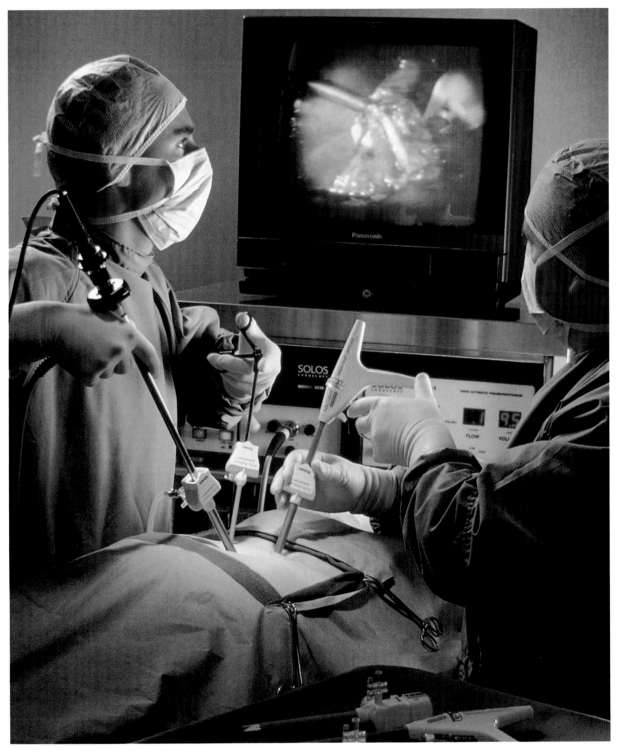

Surgeons pioneering instrumentation developed by the United States Surgical Corporation to perform minimally invasive gall bladder removal. COPYRIGHT © 1991 UNITED STATES SURGICAL CORPORATION. ALL RIGHTS RESERVED. REPRINTED WITH THE PERMISSION OF UNITED STATES SURGICAL CORPORATION.

In 1950, scientists and engineers used foot-long slide rules as portable calculators. Calculations were slow, requiring 15 to 30 seconds, and the engineer had to write down each result before performing the next calculation. The slide rules cost $30 or more, and their accuracy was good to only two or three digits. By contrast, handheld calculators of the 1990s are as small as plastic credit cards, can be powered by sunlight or lamplight, are so inexpensive that they have become throwaway items, and can calculate instantly with accuracy to six digits.

Physicians in 1950 relied on X-ray images for limited access to the body's inner workings. Even common surgery, such as repairing ligaments and cartilage in the knees, required massive invasion of the body, immobility after the surgery for six weeks or more, and months of physical therapy afterward.

In the 1990s, thanks to the combination of microprocessors, lasers, fiber optics, and other sensors, the practice of surgery has been transformed. Physicians and surgeons can observe the body's innermost functions and make diagnoses and subsequent repairs without

German air traffic controllers use highly sophisticated hardware and software to manage flight operations.

the trauma of massive invasion. Arthroscopic knee surgery, for example, involves making two minor holes in the skin. Microcameras and microinstruments are inserted, and the surgical team manipulates the instruments while observing the inside of the knee on a large video monitor. When surgery is finished, the patient is generally home the same day, walks on crutches within a couple of days, and walks normally within a week. Similar microsurgery miracles are performed inside other body parts with equally good results.

Advanced signal processing techniques and informative displays enhance the safety of air travel.

Commercial airlines in 1950 offered a two-stage flight from East to West Coast; usually New York to Chicago, then onward to Los Angeles or San Francisco. The flights were often delayed by bad weather or mechanical problems. Some 17 million persons, only one American in nine, flew on commercial airlines in 1950. Radar control was minimal,

Shrinking Calculator Thickness
1969–1979

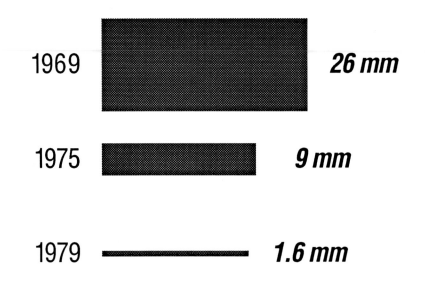

1969   **26 mm**

1975   **9 mm**

1979   **1.6 mm**

Office of Technology Assessment, from VSLI Research.

Improvements in circuit fabrication made calculators dramatically smaller and cheaper over a 10-year span.

and controllers usually knew where aircraft were only because the pilot gave an estimated position.

More than 424,000,000 persons, almost twice the population, flew as passengers on domestic airlines in 1990. Flying coast-to-coast nonstop was as commonplace as driving to the next city, and it took major storms to disrupt the orderly flow of airliners. Powerful radar systems tracked every airplane, both in the sky and on the ground.

In combat, knocking out a target such as a bridge required an average of about 500 sorties in World War II and the Korean War. In the Vietnam War, at least five sorties were required for the same job because the bombs struck far away from the aim point.

By the time of the Gulf War of 1991, one or two "smart bombs" launched by pilots from technologically advanced airplanes struck within feet of the bull's-eye. Attack missiles such as the Tomahawk flew hundreds of miles and then hit their targets with similar accuracy. Moreover, an advanced weapon such as the Patriot missile could blunt the attacks of Iraq's Scud ballistic missiles aimed at cities and bases.

Correspondents covering the Korean War in 1950 wrote their stories on portable typewriters and turned over the copy to military censors for checking and sending onward by teletype to newspapers at home. The news appeared a day to a week after the events. Photographs were transmitted by radio waves and were scarce and expensive. During the Vietnam War, television footage was airlifted from the war zone to Hawaii or Japan and then sent by cable to be broadcast on home screens.

By contrast, newsmen covering the Gulf War of 1991 used miniature video cameras and radio transmitters to flash pictures and commentary up to satellites and from there down to networks such as CNN, giving viewers anywhere on the globe immediate access to events on both sides of the war.

Weather forecasts of 1950 relied a lot on historical records and a thin network of reporting and forecasting stations scattered around the nation and the world. Forecasts were communicated via radio and newspapers and were seldom useful beyond a day or two into the future. Major phenomena such as hurricanes and typhoons slammed into populated areas with little warning, causing widespread devastation and loss of life.

In the 1990s, everyone with access to a television set can see pictures from satellites that show major global weather patterns, hurricanes as they develop, and detailed projections for up to five days ahead. The jet stream, little known and less understood in 1950, is projected over the North American continent to show patterns past, present, and future. Early warning of hurricanes such as Andrew in August 1992 enable government and people in threatened areas to take steps to minimize loss of life.

New discoveries and new technologies promise even more startling changes in the years ahead. We can only begin to imagine where we will go from here.

Driver training, accident reconstruction, and taking journeys through the mountains, deserts, or even time will be commonplace as driving simulators become more prevalent.

Chapter 12
Where Do We Go from Here?

The technological explosion continues. It will bring the eradication of many diseases; lifelong learning; broad access to information resources; and even virtual vacations anywhere on Earth or space.

I t is difficult to extrapolate from all these rapid transformations to predict the future. Recall the Royal Astronomer who said in 1956, "Space travel is utter bilge." Still, a measure of informed conjecture is possible. Alvin Toffler, author of *Future Shock*, said that the difficulties of knowing the future ought to chasten and challenge, not paralyze. Toffler reminded his readers that Francis Bacon said that "knowledge is power." Toffler believes that can be translated in contemporary terms into "knowledge is change." Certainly, it is safe to say that the new technologies of the past four decades have produced enormous change and created extraordinary additional bodies of knowledge that will lead to still more change.

Applying Technologies

Scientist James R. Johnson has described technology as "the application of knowledge, tools, and the skills to solve practical problems and extend human capabilities." It will take highly trained people to develop and apply the advances in technologies as they occur in the years ahead.

As security needs strengthen in both the public and private sectors, transponders encased in plastic cards can transmit identification codes and afford proper access to secure areas.

Currently, too few people in the work force have the basic skills needed to function in a technologically advanced society. Those who develop and apply technologies have taken that into account by designing systems that are simple to use. The user of a credit card does not need to know how the mechanism that scans the card functions or how a distant computer okays the purchase. Nor does the clerk handling the purchase need the knowledge. We don't need to understand the technologies of fax machines and satellite telephone calls. But someone somewhere along the line needs to know how things work. If not, society will malfunction when the power fails . . . as it surely will.

But there is hope.

Forecasts for the Future

Malcolm R. Currie, chairman emeritus of Hughes Aircraft, has spent most of his career either inventing the future or speculating about its shape. He has offered his own predictions for the year 2025:

- Air traffic control by satellites
- Free trade worldwide
- Push-button warfare at long distances that will make battle casualties politically unacceptable
- Computers a hundred thousand times more powerful than those of the early 1990s
- Real-time universal language translation for both voice and text, which will reduce cultural differences
- Eradication of most diseases in the world through computer power simulating biological systems and synthesizing new molecules
- Accurate global and local weather forecasts up to one month in advance
- Virtual reality progressed to the point where consumers can buy disks with 3-D databases that will allow them to experience vacations and cultural events while sitting at home
- Real-time person-to-person communications from any point in the world to any other point. Each person may have a unique personal communication number, accessible anywhere in the world.

Government and industry are already looking at the future through such projects as TravTek, a research project on intelligent vehicles and highways under way in the Orlando, Florida, area. Its aim is to explore the effectiveness of onboard driver information systems, using one hundred autos and thousands of drivers. The TravTek project integrates information from electronic sensors embedded in smart highways with information from other sources and transmits it into timely and comprehensive information systems for drivers of "smart cars." Drivers have access to systems in the cars that help them navigate, select routes, and retrieve information about the locality including data about hotels, restaurants, and events. Another possible application of these technologies might be economical electric autos linked with a convenient national network of recharging stations that would proliferate like fast-food outlets.

When one begins speculating about applications of the new technologies, there is no limit to what may be possible. But it is good to recall that things do not always work as planned and that the most sophisticated of technologies can let us down at the most awkward moments. For example, what happens when a software bug disables the telephone company's switching systems?

There are other concerns. Society may be developing into two camps, those who can cope with and use the new technologies and those who can't. That may have grave implications for societal tensions. Still, it is fair to say that the opportunities in science and technology have never been greater, and they keep expanding, thanks to the work of innovative scientists and engineers.

The past four decades of unprecedented technological progress not only help us understand the world around us better, but also provide some pointers for the future. It is clear that a single discovery can lead to scores of innovative applications. The integrated circuit,

Adventurers of all ages await the thrill of plunging over Niagara Falls in a leisure industry simulator.

TravTek Equipment in the Vehicle

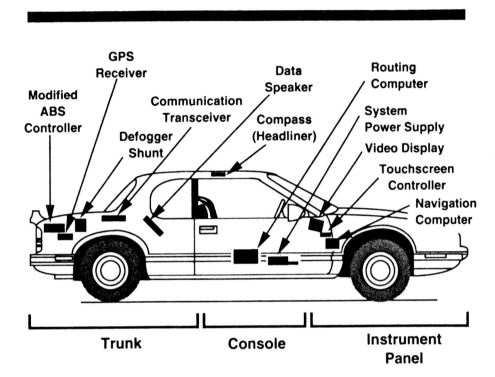

The TravTek (for Travel Technology) research program began in Orlando, Florida in 1992. It is a large-scale public and private sector project that integrates intelligent vehicles in a timely and comprehensive information network.

the microprocessor, and the laser all were important discoveries in their own right. And they spawned scores, if not hundreds, of applications once they were understood. When discoveries are harnessed together for new uses, their power is multiplied manyfold. Communications satellites and fiber optic cables provided the highways for tidal waves of information, but powerful microprocessors were required to generate and control the flow and to interpret the data when they were received. Radar and thermal-imaging sensors were able to collect masses of data. But, again, wide-band communications paths, as well as massive computing and signal processing power and memory storage, were needed to process, interpret, and display the information in useful ways.

As a consequence of the breakthroughs in electronics, people have come to expect more capabilities at lower costs from the product of new technologies. They expect instant communications worldwide at local calling prices. They expect ever more powerful computers at ever-lower prices, and get them. The recent remarkable technological advances have created strong societal expectations for technologies of the future.

We have learned, finally, that no one can predict all the eventual uses of a discovery. Who would have guessed that fiber optic cable created for long-distance communications would be equally efficient as a tool in microsurgery? Or that computing power intended for massive scientific calculations would also be suitable for turning minuscule electrical pulses into accurate pictures of the inside of a human brain?

The young men and women who are in school today are the scientists and engineers of tomorrow. They will transform the future even more profoundly than did their predecessors from 1950 to 1992. The world of 2025 that they will shape will be far different in many ways from the world we know today, just as today's world bears little resemblance to that of 1950. But getting there will be every bit as exciting and just as full of surprises as the last four decades have been.

Inventing the Future
Bibliography

Abrams, Malcolm, and Bernstein, Harriet. *More Future Stuff: Over 250 Inventions That Will Change Your Life by 2001.* New York: Penguin Books, 1991.

Adas, Michael. *Machines as the Measure of Men: Science, Technology, and Ideologies of Western Dominance.* Ithaca, New York: Cornell University Press, 1989.

Air Combat, The New Face of War, Alexandria, VA: Time-Life Books, 1990.

Allen, Thomas B., Berry, F. Clifton Jr., and Polmar, Norman. *CNN: War in the Gulf.* Atlanta: Turner Publishing, 1991.

Antebi, Elizabeth. *The Electronic Epoch.* New York: Van Nostrand Reinhold, 1982.

Berry, F. Clifton Jr., "Congestion Ahead," and "Three Minutes is an Eternity," Performance 2 (1990). Unisys Defense Systems.

_____. The Battle of 73 Easting. *National Defense* magazine (November 1991): p. 6.

_____. *Gadget Warfare.* New York: Bantam Books, 1988.

_____, et al. *America's Next Crisis: The Shortfall in Technical Manpower.* Arlington, VA: The Aerospace Education Foundation, 1989.

Brand, Stuart. *The Media Lab: Inventing the Future at MIT.* Cambridge, MA: MIT Press, 1988.

Brennan, Richard P. *Dictionary of Scientific Literacy.* New York: John Wiley and Sons, 1992.

Business Roundtable on International Competitiveness. *American Excellence in a World Economy.* New York: The Business Roundtable, 1987.

Conduct of the Persian Gulf Conflict, an interim Report by the Secretary of Defense to Congress, July 1991.

Corning Incorporated, The Next 20 Years of Fiber Optics, SR-11 Special Report issued 4/90. New York: Corning, Inc., 1990.

Crandall, Robert W. and Flamm, Kenneth, Eds. *Changing the Rules: Technological Change, International Competition, and Regulation in Communications.* Washington, DC: The Brookings Institution, 1989.

Dam, Kenneth W. *The Rules of the Game: Reform and Evolution in the International Monetary System.* Chicago: University of Chicago Press, Midway Reprint Edition, 1988.

Derian, Jean-Claude. *America's Struggle for Leadership in Technology.* Paris: Editions Albin Michel, 1990. English Translation: Cambridge, MA: MIT Press, 1990.

Dertouzos, Michael L., et al. *Made In America: Regaining the Productive Edge.* Cambridge, MA: MIT Press, 1989.

Devereux, Tony. *Messenger Gods of Battle: Radio, Radar, Sonar; The Story of Electronics in War.* London: Brassey's (UK), 1990.

Diebold, John. *The Innovators: The Discoveries, Inventions, and Breakthroughs of Our Time.* New York: Truman Talley Books, 1990.

Disston Ridge, Inc. *PC-Glossary Shareware.* St. Petersburg, FL. Disston Ridge, Inc. Software and Consulting, 1991.

Duke, David A. *A History of Optical Communications.* SR-7 Special Report. New York: Corning, Inc., 1983.

Frisbee, John L., Ed. *Makers of the U.S. Air Force: An Air Force Association Book.* Washington, DC: Pergamon-Brasseys, 1989.

Hallion, Richard P. *Test Pilots: The Frontiersmen of Flight.* Garden City, NY: Doubleday & Co., 1989.

Hellemans, Alexander and Bunch, Bryan. *The Timetables of Science: A Chronology of the Most Important People and Events in the History of Science.* New York: Touchstone Books, 1988.

Hudson, Heather E. *Communication Satellites: Their Development and Impact.* New York: The Free Press, 1990.

International Business Machines Corporation brochure. *IBM . . . Yesterday and Today.* Armonk, New York, 1985.

_____. *How One Company's Zest for Technological Innovation Helped Build the Computer Industry.* Armonk, New York, 1984. Reprinted from July-September 1984 issue of *Think* magazine.

Kearns, David T. and Doyle, Denis P. *Winning the Brain Race: A Bold Plan to Make Our Schools Competitive.* San Francisco: Institute for Contemporary Studies Press, 1988.

Keegan, John. *The Face of Battle.* New York: The Viking Press, 1976.

Kilby, Jack S. Invention of the Integrated Circuit, *IEEE Transactions on Electron Devices,* Vol. ED-23, No. 7, July 1976.

Kurzweil, Raymond. *The Age of Intelligent Machines.* Cambridge, MA: The MIT Press, 1990.

Luffsey, Walter S. *Air Traffic Control: How to Become an FAA Air Traffic Controller.* New York: Random House, 1990.

Magaziner, Ira, and Patinkin, Mark. *The Silent War: Inside the Global Business Battles Shaping Americas Future.* New York: Random House, 1989.

Mahon, Thomas. *Charged Bodies: People, Power, and Paradox in Silicon Valley.* New York: New American Library, 1985.

Margiotta, Franklin and Ralph Sanders, eds. *Technology, Strategy and National Security.* Washington, DC: National Defense University Press, 1985: GPO.

Mokyr, Joel. *The Lever of Riches: Technological Creativity and Economic Progress.* New York: Oxford University Press, 1990.

National Aeronautics and Space Administration. Scientific and Technical Information Branch. *Pioneer Venus.* Washington, DC: GPO, 1983

National Academy of Sciences. *Profiting from Innovation: The Report of the Three-Year Study from the National Academy of Engineering.* Howard, William G. Jr. and Guile, Bruce R., Eds. Washington, DC: The Free Press, 1992.

National Geographic Society. *Frontiers of Science: On the Brink of Tomorrow.* Washington, DC: National Geographic Society, 1982.

National Academy of Engineering. *10 Outstanding Achievements 1964-89.* A special brochure commissioned for the 25th anniversary of the National Academy of Engineering, December 5, 1989. Washington, DC: 1989.

Neuharth, Clark R. *Ultra-High Purity Silicon for Infrared Detectors: A Materials Perspective.* Bureau of Mines Info Circular IC 9237, 1989.

Organization for Economic Cooperation and Development, Centre for Educational Research and Innovation (CERI). *Information Technologies and Basic Learning: Reading, Writing, Science, and Mathematics.* Paris, 1987.

Paulos, John Allen. *Innumeracy: Mathematical Illiteracy and Its Consequences.* New York: Hill and Wang, 1988.

Pool, Ithiel de Sola. *Technologies Without Boundaries.* Cambridge, MA: Harvard University Press, 1990.

President's Council on Competitiveness. *Picking Up the Pace.* Washington, DC: n.d.

_____. *Competitiveness Index.* Washington, DC: Council on Competitiveness, 1989.

Puckett, Allen E. Electronics, Aeronautics, and Space. The 69th Wilbur & Orville Wright Memorial Lecture to the Royal Aeronautical Society, London, December 11, 1980.

Pugh, Emerson W., *et al. IBM's 360 and Early 370 Systems.* Cambridge, MA: The MIT Press, 1991.

Ramo, Simon. *The Business of Science: Winning and Losing in the High-Tech Age.* New York: Hill and Wang, 1988.

Rosen, Harold A., "The History of Geostationary Communications Satellites," an address on the occasion of the L. M. Ericsson Centennial Symposium, May 1976.

Ross, Ian M. *Globalization of Manufacturing: Implications for U.S. Competitiveness.* Statement before the Subcommittee on Technology and Competitiveness, Committee on Science, Space and Technology. U.S. House of Representatives, October 3, 1991.

Saunders, Robert J., *et al. Telecommunications and Economic Development.* Baltimore: World Bank Publication: Johns Hopkins University Press, 1983.

Smith, George S. The Early Laser Years at Hughes Aircraft Company. *IEEE Journal of Quantum Electronics*, Vol. QE-20, No. 6 (June 1984): p. 577.

Stimson, George W. *Introduction to Airborne Radar.* El Segundo, CA: Hughes Aircraft Company, 1983.

Tarassuk, Leonid, and Blair, Claude. *The Complete Encyclopedia of Arms & Weapons.* New York. Simon and Shuster, 1982. English language translation of the 1979 original in Italian, published by Arnoldo Mondadori Editore.

Thurman, Richard A. and Mattoon, Joseph S., "Virtual Reality: Theoretical and Practical Implications." Proceedings of the 13th Interservice/Industry Training Systems Conference, December 1991, pp. 86-95.

Truxal, John G. *The Age of Electronic Messages.* Third Printing. Cambridge, MA: The MIT Press, 1991.

U.S. Congress. Office of Technology Assessment. *Critical Connections: Communication for the Future.* Washington, DC: GPO, 1990.

_____. Office of Technology Assessment. *Rural America at the Crossroads: Networking for the Future.* Washington, DC: GPO, 1991.

————. Office of Technology Assessment. *Miniaturization Technologies.* Washington, DC: (OTA-TCT-514) GPO, 1990.

————. Office of Technology Assessment. *Helping America Compete (The Role of Federal Scientific & Technical Information).* U.S. Congress, Washington, DC: 1990.

————. Office of Technology Assessment. *Information Technology and R&D: Critical Trends and Issues.* Washington, DC: (OTA-CIT-268) GPO, 1985.

U.S. Department of Commerce. *The Story of the U.S. Patent and Trademark Office.* Washington, DC: GPO, 1988.

————. United States Telecommunications in a Global Economy: Competitiveness at a Crossroads. Washington, DC: GPO, August 1990.

————. Technology Administration. *Emerging Technologies: A Survey of Technical and Economic Opportunities.* Washington, DC: GPO, Spring 1990.

————. *Statistical Abstract of the United States.* Washington, DC: GPO, 1991.

————. *U.S. Industrial Outlook 92: Business Forecasts for 350 Industries.* Washington, DC: U.S. Department of Commerce, International Trade Administration, 1992.

————. *Emerging Technologies: A Survey of Technical and Economic Opportunities.* Washington, DC: Spring 1990.

————. *The Competitive Status of the United States Electronics Sector from Materials to Systems.* Washington, DC: GPO, April 1990.

U.S. Department of the Interior. Bureau of Mines. *Mineral Commodity Summaries.* Washington, DC: January 1991.

United States Patent Office. Patent No. 3,758,051: awarded September 11, 1973 to Donald D. Williams.

————. Patent No. 3,138,743: awarded June 23, 1964 to J. S. Kilby.

Van Creveld, Martin. *Technology and War: From 2000 B.C. to the Present.* London: Brassey's (UK), 1991.

Wellman, David A. *A Chip in the Curtain: Computer Technology in the Soviet Union.* National Defense University Press: Washington, DC: GPO, 1989.

Westinghouse Electric Corporation. *Centennial Edition of* Engineer *[Magazine].* Pittsburgh, PA: Westinghouse Electric Corporation, July 1986.

Williams, Frederick. *The New Telecommunications: Infrastructure for the Information Age.* New York: The Free Press, 1991.

Index